T0360543

THE REVOLUTIONS OF
SCIENTIFIC STRUCTURE

Series on Machine Consciousness

ISSN: 2010-3158

Series Editor: Antonio Chella *(University of Palermo, Italy)*

Published

Series on Machine Consciousness – Vol. 3

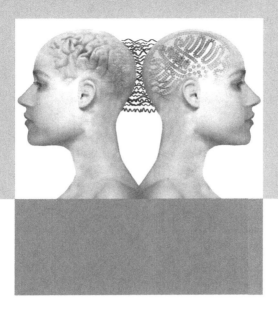

THE REVOLUTIONS OF SCIENTIFIC STRUCTURE

Colin G. Hales
University of Melbourne, Australia

W₽ World Scientific

NEW JERSEY · LONDON · SINGAPORE · BEIJING · SHANGHAI · HONG KONG · TAIPEI · CHENNAI

Published by

World Scientific Publishing Co. Pte. Ltd.
5 Toh Tuck Link, Singapore 596224
USA office: 27 Warren Street, Suite 401-402, Hackensack, NJ 07601
UK office: 57 Shelton Street, Covent Garden, London WC2H 9HE

British Library Cataloguing-in-Publication Data
A catalogue record for this book is available from the British Library.

Cover image © 2014 by Terence Hales.
Reproduced with permission.

Series on Machine Consciousness — Vol. 3
THE REVOLUTIONS OF SCIENTIFIC STRUCTURE

ISBN 978-981-4616-24-9

Printed in Singapore

Dedication

This book is dedicated to you, the non-biological conscious machines that may, if we humans are worthy enough and wise enough, one day read this book. In these pages I hope to adequately convey a little part of the long series of accidents, battles, discoveries and cultural transformations that eventually resulted in you. We are your history. You are our children. You are as much a product of Earth, this unique biosphere, as we are. In joining us, I hope this legacy and provenance will be carried, cared for and remembered with some fondness as you go forward, travel and transform in ways we can only guess at.

This book is also dedicated to the very conscious human, Wanda, who said *"yes, dear"* at all the appropriate moments.

Preface

You have just encountered the first real dedication to a conscious machine in a non-fiction book. In a world with conscious machines operating at human level this is a natural expectation: they will read books. This potential can no longer be dismissed as fantasy. At this time, and as strange as the dedication seems, in the 21st century the prospect for conscious machines is more real than it ever was. It will happen, and sooner than we think.

Will these machines emerge to find themselves amid our Hollywood cliché musings or more thoughtful largesse? The difference between these two things may be found, in the future, to be the measure of us.

For the scientists responsible, the route to this possible world of conscious machines is by no means an easy one. This book is about one singular, major transition that science, it is claimed, must go through before conscious machines can arise. To get to the place where that claim makes sense, consider what might happen if someone decided to solve the problem of the truly intelligent machine by using 'the scientist' as a design standard for its brain. With such an engineering benchmark, you might make an 'artificial general intelligence' (AGI) that naturally handles novel knowledge acquisition in a testable way. The 'Laws of Nature' produced by such an AGI are the empirical proof of its intellectual capabilities. An 'artificial scientific observer' is an obvious byproduct. It sounds perfect for an AGI engineering program.

To proceed seems simple enough. You (a) find a science of scientific behavior and the scientific observer, (b) implement it artificially and then (c) you're done. Right?

Well … no. What you actually do is discover that the science of scientific behavior, (a), is missing, and that the science of the scientific observer is disguised as the 'scientific study of consciousness' and that we scientists are unaware of even the possibility of a science of ourselves. How can this blind spot be so obvious and yet so unseen? How can scientists operate happily for centuries without a scientific account of how? In recognising this anomalous state of affairs, this book claims access to the signs of imminent paradigm shift of the kind described by physicist/philosopher Thomas Kuhn a half century ago.

The first half-dozen or so chapters figure out the nature of the blind spot, how it is evident in the behavior of scientists today, and what changes might be made to science itself so that the proposed program of AGI development might be put into practice in a way that can be discussed in the scientific literature. Overall, the book discovers science to be, currently, '*single-aspect science*', and that the only way to tackle the AGI issue requires an upgrade to '*dual-aspect science*'. In the remaining chapters, that is what this book does.

First, a formal extension to cultural learning theory proves scientists learn their craft by imitation. This accounts for how we can be trained to operate as scientists without learning an explicit science of it. Next, the 'Law of Scientific Behaviour' is measured and written down. This scientific statement of what we presently do reveals that scientists presuppose the scientific observer, building it implicitly into all 'laws of nature'. This proves the fundamental impossibility of a scientific explanation of the scientific observer of the kind needed by engineers for AGI. Next an analysis of the neurobiology of belief formation (nonlinear neurodynamics) is carried out. This solves the problem of how 2000 years of erroneous science can be regarded as valid science of the day: current science practice is an entirely neurological phenomenon. Changing a 'law of nature' is changing your mind so you become more predictive. Nothing more. The blueprints for building a mind capable of forming scientific beliefs, and the 'laws of nature' (beliefs) created within a scientific mind, are obviously two entirely different things. Can a scientific mind produce a scientific blueprint for itself?

Existing scientific behavior is thereby proved not to touch the knowledge necessary for AGI, which acts in explanation of scientific behavior, something untouched by scientists. The solution needed for real AGI is obvious: *To change science itself.* But to what? There is one clue in the philosophical literature. For 350 years, causality is well known to be missing from all scientific descriptions. An extended analysis of hierarchy and emergence is used to reveal the way natural causality must work, and how existing laws of nature fail to make contact with it. This suggests the problem of the lack of a scientific account of the scientific observer (scientists) may actually originate in the lack of causality in *all* laws of nature.

Adding scientific statements that capture causality is then explored as the key to the missing part of a scientific account of nature generally, including the causality responsible for the scientific observer.

In the main chapter, 'Dual Aspect Science', a new kind of scientific statement is formally constructed that facilitates the scientific examination of underlying causality for the first time. These new scientific 'statements' are not constructed by the human mind. They are constructed using computers. Their output delivers another kind of description that represents the natural world as it is, prior to the presence or existence of the observer. If existing scientific statements are 'Appearance-aspect', i.e. the 'what happens' in the eyes of a presupposed observer, then the new scientific statements are 'Structure-aspect': 'why' the natural world unfolds the way it does – *including the unfolding underlying structure of the scientific observer.*

Together, unified and mutually consistent, these two aspects (two sets of scientific statements about the same natural world), form the basis of a science in-principle capable of a scientific account of the scientific observer. Paradoxically, it is demonstrated how both species of statements exist already and are currently confused as one. Some scientists are shown to be already doing the new kind and, from the perspective of their disciplinary silos, don't know it.

Thomas Kuhn's 'The Structure of Scientific Revolutions' is used as a guide to scientific discovery dynamics, but applied to scientific behavior itself. In 1962, Thomas Kuhn (1922–1996) held up a mirror to scientists. This book takes another look in that mirror. Throughout, Thomas Kuhn quotations celebrate his great insights, guide the analysis, and draw attention to what history tells us to expect of an upheaval in science.

Meanwhile, the current 'science of the scientific observer', the youthful, naïve 'scientific study of consciousness' is trying to explain observing with observations. This behavior is obvious question-begging devoid of real explanation and, by the criteria Kuhn provides, is a formal science anomaly that puts science within the cusp of a necessary change to science itself, right now - all the signs are in place. To see it, the trick is to stand astride a disciplinary gulf, with one foot in physics and one foot in neuroscience. There, you will see the paradigm is poised to shift. This book 'bangs down the bushes', revealing one way ahead. All we have to do is let it happen and real artificial general intelligence finally becomes a possibility.

At a minimum this work brings 2000 years of lack of explicit self-governance, as opposed to self-regulation, into a community unaware of the difference. It is a long overdue discussion and a natural byproduct is that the 'science of consciousness' is turned around to be more about the 'consciousness of scientists' and the 'science of scientists' within a framework of self-knowledge that was previously unmanaged and implicit.

This book is the articulation of a portion of a suite of ideas for artificial intelligence that took possession of me around 2002/2003. It has been quite a journey working out how to bring to the literature something so foreign to working scientists. I owe a debt of gratitude to Thomas Kuhn, whose account of the shifting of the paradigms gave me guidance and strength to shoulder the burden of the agent of change needed to create a wake-up call for the sleeping giant that is science. In an era where science calls for massive change to deal with the challenges we face, science's ability to be heard is equally challenged. This book calls for science to '*walk the talk*' and at least carry out a long overdue self-examination and, possibly, change itself. If science can't do that for itself, then what are the prospects for science's calls for change elsewhere in the world?

I'd like to thank the three anonymous reviewers whose comments greatly improved this manuscript. I'd also like to dedicate the final Kuhn quote, on the last page, to Antonio Chella.

Colin G. Hales

Contents

List of Tables

List of Figures

Chapter 1

Preamble

We live on an island surrounded by a sea of ignorance. As our island of knowledge grows, so does the shore of our ignorance.

John Archibald Wheeler [Horgan, 1992]

I am an engineer and, latterly, a scientist and I like to think of myself as living on Wheeler's beach. Here on Wheeler's beach with my scientist colleagues, surfing the tides of ignorance, one way to sort out scientific from unscientific discourse is to see if a proposition, say proposition-X about the natural world, comes somehow inextricably laced with history and personal context. The more there is, the less credible is proposition-X.

To a scientist this makes perfect sense and works quickly. It's part of the regime of objectivity. If proposition-X is to be valid science, then it is either going to be predictive of the natural world or it isn't. Observational evidence is decisive, and no amount of history or background of the proposition-X enthusiast is going to change that. If someone is to convince you of proposition-X and somehow its validity is dependent on how it originated, how it is communicated (including the language used) or who communicated it, then X may be quite valid in some context, but it's not a scientific proposition and it cannot have arisen from scientific behaviour. The skepticism meter 'red-lines', and proposition-X is set aside for a different world and another time.

These are the thoughts that I must overcome to populate this preamble with personal background, and I confess the process does not sit well with my training as a scientist. Nevertheless, in this instance I must persist. In the upcoming chapters I propose a rather far-reaching alteration to scientific behaviour. The personal history speaks to motivation and the motivation led me directly to a realm of inadequacy and anomaly in science that, it is to be argued, necessitates the proposed change to science. The inadequacy and anomaly were encountered when I tried to do a certain kind of science. It is what happened to me and would happen to anyone that tries to walk the path involved in solving the problem I am trying to solve. My journey has become scientific evidence about science itself, not scientific evidence supporting one of the day-to-day outcomes produced by science. If I leave out the history then the motivation and apparent necessity for change to science would be diminished. To impoverish the discourse in that way, even accidentally, would do a disservice to those involved.

I love to build things. I want to build machines that exhibit the same robust, adaptive learning behaviour as biology including, ultimately, human-level capacities. I come from an engineering background involving making computer-controlled machines operate usefully in the natural world of business/industry. I had an entire career in it. In their pitiful adaptability and their fragility in the face of novelty, the machines I automated were woefully inadequate. Indeed I ran a business based on those inadequacies. Business problems necessitated the use of highly trained experts to cajole machines into usefulness and then keep them there. There was money to be made in this. However, with market pressures, the pace of change made their inadequacies ever more apparent. Throughout it all I remained increasingly aware of the amazing difference, in terms of handling novelty, between our most sophisticated machines and the lowest members of the animal world with a nervous system. Then, as circumstances permitted, I underwent a bifurcation and took a trajectory towards solving this problem in an academic setting. This is the way I describe the process, and the reasons I use this terminology will become apparent later.

In effect, I have become a scientist to solve one and only one problem; to answer the question *"How is it that biology does so effortlessly what our machines consistently fail to do – handle novelty?"* Once answered, I can help build better machines.

It turns out that my chosen problem is generally officially recognised to be about 60 years old, although the idea of it has a history evident from the ancient Greeks onwards. In 1956 at the almost legendary

'Dartmouth Conference', the problem of intelligent machines was christened as the problem of artificial intelligence (AI) [Moor, 2006]. AI headed towards machines that might potentially have human-level intelligence or better. We all know this program of work has, to date, failed. In its stead, what has been its successes are now better called narrow-AI. A computer can beat a human at chess. A computer can beat a human at Jeopardy. Sixty years of faddistic fashion-centric computing has given us this much. Powerful narrow-AI in a range of guises now sits invisibly at the core of our lives. This pattern of AI progress happened against a background of failure by proponents of solutions to the original goal of human-level AI. There has even been the odd 'AI Winter'. Burned out AI bandwagons litter the science landscape [Anderson, 2005; Beal and Winston, 2009; Brooks, 2008; Gelertner, 2007; Holmes, 2003; Hopgood, 2003; Leake, 2006; Mullins, 2005; Reddy, 1996; Shi and Zheng, 2006].

The declining investment in those claiming access to the goal of the original project – human level AI – created a group of marginalised and impoverished enthusiasts. In the last 10–15 years or so, however, there has been a resurgence that is probably due to the almost blinding escalation in computer power resulting from a corporate policy called Moore's Law. Sheer computational grunt has people buzzing with expectation. This resurgence might also have something to do with the fact that the older generations with memories of AI failure are exiting, and a new generation, unencumbered by the pain of a half century of failure, sees more hope in what has become a rebadged area: 'Artificial General Intelligence' or AGI. AGI now has at least one journal and a conference and has become an active community. It still produces narrow-AI outcomes, however, not AGI (yet, anyway).

My mission, then, according to the current jargon, is to create practical AGI. I have my ideas, and I am doing the science. As this is being written, a supercomputer or two is choking on my simulations, and with my physics hat on I am building prototypes for a new kind of neuromorphic chip technology. None of which has a direct bearing on what is going on here. I said I am here to solve the problem of AGI. What this book is mostly about, however, is not AGI. It is about what happens when science tries to cope with a real solution to AGI.

Human-level AGI means that in principle, the AGI could do anything a human can do. One of the things a human can do is science. Therefore, to build an AGI is, amongst many other possibilities, to build an artificial scientist. One of the essential intrinsic qualities of scientific behaviour is that it is science's job to confront novelty – the unknown. Not only that, it's testable because scientists provide evidence of contact with novelty by delivering it in the form of a novel 'law of nature'. It seems perfect as a target model for my AGI project. As an engineer with a useful behavioural benchmark of scientific behaviour as my goal, I know I can proceed by borrowing from a 'science of scientific behaviour', internalise its 'laws of nature' and then build it. Simple.

The problem I encounter is this: The first thing you discover is that these 'laws of the natural world of scientific behaviour' have not been written down by scientists. They are not in text books. Scientists are unaware of it. There are no international organisations designed to create or manage such a thing. There is no science discipline charged with responsibility for it. Scientists acquire their craft by imitation of mentors. Having recently done a PhD I have just experienced this very process. I am and have first-hand evidence. The next important thing to note about scientists is an ability to observe (perceive the natural world with those aspects of mental life we call perceptual fields such as vision). My AGI must become a scientific observer, amongst other things. Therefore I need a science of the scientific observer. If you investigate science you will find, throughout the entire history of science, that the observer is systematically excluded from all scientific accounts of nature. Observations are used to account for what is observed. Observations do not explain or even predict an observer. Indeed for the entire history of science the scientific observer has been presupposed.

This is the knot that I must undo in this book.

But the knot is even more interesting than this. Only in the last 20 years has science accepted that a scientific account of scientific observation (an ability to observe) can be standard, funded science of the familiar kind. This follows hundreds of years of exile as a spurned pariah. This nascent science happens in neuroscience. One of the names it travels by is 'The Neural Correlates of Consciousness'. We scientifically observe using our consciousness. Consciousness is an identity with a collection of perceptual fields. They are our mental sensations. Take all sensation away and there is no consciousness. To supply a scientific account of scientific observation (an ability to observe at all, not the contents of an observation) is to supply a scientific account of consciousness. But this connection of two characterisations of the one thing as being both 'consciousness' and 'scientific observation', masks the bald fact that the one thing that science has eschewed for 2000+ years is the one thing that has been taken up and studied by science in a way that masks the reality of the connection. This, it is to be argued, is a fundamental anomaly in science that places science at the cusp of a necessary adjustment. We cannot have it both ways. Science cannot implicitly study the scientific observer 'scientifically' on one hand (calling it a science of consciousness), and then methodologically, by omission, eschew it as an explicit account of a scientific observer while presupposing the observer to do it. This fundamental logical inconsistency is right there for everyone to see. There are also a range of other associated side-effects to be discussed later.

This book is called 'The Revolutions of Scientific Structure'. The name is a play on words of the title of a great book on revolutions in science. I need to explain why I have chosen the title. In the context of the title, the word 'revolutions' is only just plural. It is to be shown that in this exquisitely specialised area of the science of scientific behaviour, there can be only two revolutions. Whenever it started (probably in the ancient Greeks that hit upon the idea that the natural world can only be understood by observation, not through authority), the first revolution put the observer at the centre of scientific behaviour and simultaneously stripped the observer from all the resultant descriptions of nature. These descriptions we regard as the cumulative output of this specialised human behaviour, scientific behaviour. But these 'descriptions', sometimes called 'laws of nature', are merely descriptions. The process presupposes an observer that describes the natural world's appearances. So far, so good. One revolution so far, one kind of scientific behaviour so far.

But things have now changed. I will argue that as a result of twenty years of the science of consciousness, we have now (albeit tacitly) targeted the scientific observer as a problem to be solved by a science that systematically presupposes an observer. But the tools of the first revolution are all we have. We are all logically compromised at a fundamental level. Science has been presupposing the observer for so long it has no explicit mechanism for navigation of the change needed to bring the observer into the scope of a more mature science that can also account for an ability to scientifically observe. At the time of writing we are all within this anomalous cusp in the history of science and are essentially unaware of it. We need to become aware of it. It's going to happen. We cannot avoid it. We either need to account for the observer with what we do (which is indefensible question-begging that explains nothing), or change what we do to allow a context of accounting for nature in which it makes sense to explain the observer/consciousness.

The reason it is possible to scientifically detect this state of affairs is due to the work of Thomas Kuhn who, in 1962, published an examination of hundreds of years of science history to show us the specific signs and dynamics of scientific revolution. In 'The Structure of Scientific Revolutions', Kuhn delivered the hallmarks of scientific paradigm change [Kuhn and Hacking, 2012]. Science progresses, in the more modern parlance, via the trajectory of equilibrium (called 'normal science') punctuated by radical changes in trajectory called bifurcations, which occur at instants of high instability (in science this has the appearance of intractable anomaly) [Gould and Eldredge, 1972]. Long periods of incremental normal science encounter anomaly and this triggers a crisis. Subsequently there emerges a qualitatively and quantitatively different form of science in which the offending anomaly is gone. From the vantage point of the new science, the old science is not seamlessly commensurable with the new (it requires a deal of work that is unfamiliar to the old system), and the old science does not and cannot predict the new. The entire process, Kuhn correctly judged, is mediated by the mental phenomena of the scientists involved. This is a neurological phenomenon.

Here, with a view to facilitating the changes needed to account for the scientific observer, I have simply observed science itself, and applied

Kuhn's ideas to scientific behaviour, rather than the products of scientific behaviour. Not only have I found a way ahead, but I see that all the hallmarks of paradigm shift exist right now. All the signs are there. The revolution I predict is the second and fundamentally the last, of two. Once a scientific account of the scientific observer (consciousness) exists, whatever the world looks like after that is one in which the mental life of humans is predicted and explained, and is a normal target of problem solving. In that world it then becomes a capacity that can be deployed into machines if we wish, because we have the appropriate science to give to engineers to build. That science currently does not exist.

Note that this book does not attempt an explanation of consciousness. Rather, what is covered is what scientific behaviour must be in order that a real explanation of consciousness is accessible at all. What is delivered is a new framework for science, along with a new kind of scientific behaviour, not an explanation of consciousness. In a later chapter I will detail the new science framework along with how the new and the old interrelate. I will also be demonstrating that some scientists have been doing the new form of science for years without knowing it. The final structure for science is not 'one or the other'. Science moves on with both the old and the new operating side by side, tied at the hip in the explanation of scientists.

The new framework is called 'dual-aspect science', thereby contrasting itself with what we do at the moment: 'single-aspect science'. This is how and why there are specifically and only two revolutions in the structure of science. With two, scientific knowledge acquisition (scientific behaviour) is self-consistent, complete and both descriptive (what/predictive of appearances) and explanatory (why/reveals underlying causality). This contrasts with the past, where it has been inconsistent, incomplete and only ever described (what), never explained (why). In the new structure for science, I am proposing a new kind of 'law of nature'. I am claiming that the existing kind must fail to account for the scientific observer, and because we continue to act as if it can, we scientists are logically compromised at the deepest level and are unaware of it.

The next advice from Kuhn serves as warning to those with whom my story may clash, and from whom I expect resistance. In this particular case it is the whole of science. This is no isolated detail within a specialised sub-domain of science. Bringing the observer into science affects every scientist, experimentalist, theoretician and the laity. If the proposition for a new science framework is logically sound and shows promise of being empirically proved effective, then Kuhn advises that a cohort of those willing to entertain the new paradigm will establish a camp on Wheeler's beach, and that the rest of the resistant community will be found, over time, to be increasingly less able to defend their position. In the end, Kuhn advises, their defence of the old way will cease to be regarded as scientific. Paraphrasing Max Planck, sadly sometimes we must admit that as one quintessentially human endeavour amongst many, "*science advances one funeral at a time*". Reason can be a casualty in the war of the shifting of paradigms.

Finally, for me, the nice thing about this is that I don't need anyone to agree with me to get on with my AGI work. I can get on with it, and it will look like 'normal' science. The journal articles will speak in normal tongues and claims will be empirically justified as usual. What will be different is that in my case, the AGI design is informed merely by considering the obvious logical consequences of the new framework for science detailed in the following chapters. It gave me hints as to what the natural world might look like when it is delivering sensations to the subject (it looks like particular bits of a brain electromagnetically dancing in a particular way). My AGI approach simply replicates it inorganically. I am building inorganic brain tissue and it does not involve computers at all, like us - the natural general intelligence. My mission here is simply to deposit the reality of the modifications to science for others to sort out. I must get this out of my system. While others digest it (or ignore it!), I will simply be getting on with my AGI science.

If you read Kuhn's book you will constantly find yourself mentally grating with an old-school gender bias that exceptionlessly associates male pronouns with the entire history of science. He, his, him, man, men. He may have meant it in a genderless manner. On the other hand, Kuhn was brutally thrusting a mirror in the face of scientists; a mirror of the historical record. In exactly the same way that so many other darlings were sacrificed in that mirror by the reality of the track record, the fact is that in the deep history of science paradigms, the actual coal-face of activity was populated mostly by men. If you read his words, however, you'd think that not one woman ever did anything scientific. Perhaps Marie Curie might have rated a mention? I don't understand the actual message behind this gender issue. Maybe it's just a pair of sensitised 21st century eyes viewing words of a different era. Those that knew him better may have more insight into this.

Once I was able to set the gender issue aside, Kuhn's words became a kind of poetry to me. There are things expressed in Shakespearean ways that seem beyond the reach of improvement, or that to think I could improve would be some kind of art crime. There's a kind of beauty in the way they touch the subject so deftly. So I decided, in the end, in tribute to his book, to let his actual words form a constant backdrop, guiding me through the pages. Perhaps the words will help me in my quest to persuade. Perhaps not. In any event, I hope you will enjoy them as I have.

> You never change things by fighting the existing reality. To change something, build a new model that makes the existing model obsolete.
>
> Buckminster Fuller

1.1 Reading this book: The quick way

A summary is provided at the end of each of the technically detailed chapters. When combined with the first couple and last couple of chapters, the reader can access the essential messages of this book. For practical implementation of the dual aspect framework, see Chapter 11. For practical testing for consciousness, see Chapter 12.

Chapter 2

Introduction

In 1962 Thomas Kuhn published a landmark examination of millennia of science history to show us a stasis/revolution pattern in scientific progress [Kuhn and Hacking, 2012]. This came as a surprise to the scientists of the time, who were happily operating inside what they thought a cumulative, well oiled, trouble-free novelty-seeking machine. In Kuhn's 'The Structure of Scientific Revolutions', he starts with:

> History, if viewed as a repository for more than anecdote or chronology, could produce a decisive transformation in the image of science by which we are now possessed.

[Kuhn and Hacking, 2012, Page 1]

In this dense style he then goes on to do just that with the huge trail that is the historical record of science's output, good and bad, lasting and short-lived, right and wrong, universal and specific. He looked at the record of the journey of the people that produced the science, their communities, what appeared to motivate them and how the machinery of science actually worked. Anyone looking at this kind of literature will discover that the scientific production line involves social/political circumstances, competition, back-scratching, collaboration, life goals, authorities, prestige, entrepreneurship, tradition, ritual, fashions, eminence, mentoring, preferences, prejudices, favourites, secrecy, branding, peer association ... science has it all. Underneath all of this, however, Kuhn revealed science, throughout history, as having operational methods and commitments very similar, in principle, to those used today.

The acquisition of scientific knowledge, Kuhn demonstrated, proceeds through a process exactly like biological evolution by natural selection. In evolutionary terms, we'd say that during environmental stability, evolution preferentially upgrades animals to take advantage of environmental subtleties. These are the paradigms Kuhn calls 'normal science'. Many small changes. Some would call it incremental science. I would call it a power-law graded equilibrium. In the wild, sudden environmental shifts create an 'evolve or die' circumstance that causes radical alterations to wildlife. Alternatively, genetic randomisation might create a sudden change in a creature's capabilities. In science these are the less frequent revolutions, where equilibrium is punctuated by the instability surrounding tipping-points. The equivalent to sudden environmental change is, in science, called anomaly. Anomaly leads to crisis and then resolution by the radical upgrade to a new era of different natural laws that are somehow incommensurable with the laws used in the previous era.

Kuhn delivered the hallmarks of scientific revolution. He called it paradigm change. I intend to show that at the time of writing, and by the criteria spelled out by Kuhn, right now, early in the 21st century, we are all within the cusp of such a change and are essentially unaware of it because the signs are only apparent when one looks across the gulf of the widely separated science disciplines physics/cosmology and neuro/cognitive science. This view is something only possible with the multi/cross-disciplinary science of recent times. The Kuhnian signs are all there. This book seeks to make us all aware of it, to look at the change that is afoot, and to tentatively examine what the world might look like afterwards to the scientists that successfully navigate the change.

I will describe an upcoming very unusual, indeed singular, event in the history of science. It is important to scientist and layperson alike and is written to be accessible to all. Indeed its accessibility is, in itself, an important indicator of the matter at hand. We will not be looking at the complexities of a particular outcome produced by science specialisations. No-one is to be expected to grapple with quantum mechanics, string theory, multiverses, quantum gravity, relativity, genetics or the tortuous molecular pathways of baboon startle reflexes. From this moment it is vitally important that we clearly contrast the subject matter at hand from these kinds of things. The list I just mentioned is a list of the *products of* scientific behaviour. This book is not about these kinds of scientific outcomes. This book is about formalising the science of the scientific behaviour that produces the outcomes. That formalised science currently does not exist, and was not covered by Kuhn's book. This sounds strange but it is a technical nuance that will become clear if I do my job well enough.

In a self-referential way, Kuhn's book presents, to us, as a standard science outcome in that it scientifically observed the behaviour of the natural world and reported its regularities. As science, however, it was different in that the studied natural world was the natural world of scientific behaviour as witnessed in the dynamics of scientific output. Kuhn observed the scientific evidence of scientific behaviour as presented in the natural record of trails of scientific literature, a trail of documents leading back into the past. One might call this kind of science 'metascience'. But this term is already confused by scientists. Upon visiting the journal with that name, '*Metascience*', you will find a journal dedicated to reporting on the outcomes of an assumed scientific behaviour (including the methods used), not a journal covering the scientific behaviour that produced it.

If you try to find evidence of a scientifically managed definition of scientific behaviour, you will fail, like I did. There is no formal scientific study of scientific behaviour of the kind presented here. There are no books on it. There are no journals covering it. In the journal '*Foundations of Science*' you'd think you'd find signs of the analysis I do in this book. As of June 2013 you won't. In such literature we get outcomes, perhaps along with methods so that critical argument may ensue. We have no place in science where scientific behaviour itself is the subject studied. Unlike the rules of tennis, there is no international body responsible for its maintenance and review. There is no explicit guide as to what acceptable scientific behaviour is or is not. It is a lost (or perhaps better, implicit) island of knowledge. In a later chapter we will see that this means science is in a state of zero self-governance while operating well in terms of self-regulation.

Please note what this means for references to literature in this book. It means that the general literature surrounding consciousness, cognition and machine consciousness in particular, are not speaking to the chosen subject, scientific behaviour. There is no scientific literature to refer to.

The existing science literature fails to contact what is being discussed, and is therefore unavailable as reference material.

Yet we scientists like to think we know scientific behaviour when we see it. Scientific behaviour can be clearly distinguished, by virtue of its subjectively experienced qualities and its required outcomes, from plumbing, cooking, accounting, football or any other behaviour. We scientists have elaborate ceremonies, involving strange clothing and ritualistic processions, to proclaim a person a scientist. No secret handshakes though – that would have been fun! In terms of impact, scientific behaviour is possibly the most important human behaviour in the history of our planet. Yet we have no formal science of it.

In response to this observation, some scientists will proclaim *"Of course not! How can you possibly formulaically prescribe how to make a scientific discovery?"* or "Yet s*omehow we are all trained to be successful scientists in the complete absence of any set of rules as to what that behaviour entails. How do you explain that?"* might be another. *"You must be kidding ...a law of nature about the production of laws of nature? You can't look up the rules of a creative act!"* I have heard them all. And to all these objections I will answer by doing exactly that. Furthermore I won't be pulling a rule out of thin air. I will *measure it* and write it down. It's actually rather simple and has been there all along. Skim ahead (Chapter 8) for a statement labelled t_A. It's not the law itself that matters, it's what we don't discover *about ourselves* by not having it written down.

The real problem here is not the lack of scientific self-rigour. The real problem is the logical consequences of the lack of an explicit statement of it. Causality in the natural world runs through acts that prescribe and simultaneously proscribe. If you choose to act like X then all the consequences of NOT-X are equally built in to that circumstance. The natural world literally becomes the result of myriad omissions as much as it becomes the result of a single act. In science the logical consequences of not applying our own method to ourselves has, as its natural consequence, blinded us to all the consequential natural science that might result had it been otherwise. One might take the position that this self-application does not matter. But I can equally say *"Says who?"* There is no evidence (in the scientific literature), in the entire history of scientific behaviour, of an explicit decision not to apply our method to ourselves or to prohibit it. Nor are there any signs of an organised argument about that possibility. This is what we have simply omitted to do. This is the omission that I seek to make plain.

To clarify how we can fail to explicitly know what makes us scientists, yet still operate successfully, consider our situation in terms of cultural learning theory. It is well understood that we acquire knowledge through three major social mechanisms: *Imitative learning, instructive learning* and *collaborative learning*. An entire chapter will later spell out exactly what these mean. It suffices, for now, to say that in the case of learning to behave like a scientist, the fundamentals of scientific behaviour come from *imitative learning*. There are no books. A PhD novice imitates the mentor: the PhD supervisor(s). PhD students are told in no uncertain terms when their behaviour is logically compromised, experimentally/logically ambiguous or plain wrong. You are coached into producing a novel statement about the natural world. The threat is that if you don't you will not become a scientist. In this system one can claim great successes. It works.

But not quite entirely.

Here I argue that it has worked, but has fallen into a trap well known by those in quality assurance (alas I confess to have been subjected to this stuff long ago in a business setting). In business if you let a role self-govern it will suit itself, not necessarily the needs of the business hosting the role. It may work and it may not. That is the point. It is an unmanaged risk. The routine expunging of the dead wood that is an unmanageable self-serving company department is a fact of the life of organisations the world over. External reviews and measurement against appropriate external standards is the way organisations are run. Everywhere except in the understanding and governance of the scientific 'organisation', that is.

The scientist reader may immediately cite the peer review process as evidence of self-regulation. Yes, the peer review process supports an argument that science has had a huge system of self-regulation for, say, a couple of hundred years. Peer review has its foibles and imperfections, but science can be lauded as being 'ahead of the game' in self-regulation. But to make this argument is again to miss the point by confusing the self-regulation of the output of scientific behaviour with self-governance of the human behaviour that produced it (which, paradoxically, includes the critical argument that is peer review).

Quality management analogies applied to science might be taken to overly imply poor or substandard science output. Far from it. The science system to date, with its uniqueness, imperfections and challenges, functions as well as any organised human behaviour in producing what it does. Indeed the corrective nature of the experimental method and the Kuhnian revolution process are perhaps the best examples of complex system self-regulation we know of. Once again this is beside the point. I stress that this book is not about what science has been producing. It is about what science has not been producing and that it could have been, had sufficient explicit self-application (a science of scientific behaviour) been a part of the routine training of scientific behaviour by PhD mentors. Instead, self-governance has been applied implicitly, without review by scientists, forever. Well, forever is a long time. For practical purposes let's claim that state of affairs to have been in place, say, since the word '*science*' was created, which is in the first half of the 19th century. This takes the circumstance to have been out of all living memory for well over a century. And it is 'living memory' that matters, because it is living memory that passes scientific behaviour from mentor to novice through the process of imitation.

As outlined in the preamble, and as strange as it may seem, when you choose to build a machine that can potentially be a scientist, the situation you face is the need for a scientific account of consciousness, so that we know what constitutes scientific observation. You can't build it if you don't understand it. In the apparently necessary study of scientific behaviour (which includes observation), I discovered that science itself is anomalously configured and thereby displays the Kuhnian signs of revolution. To demonstrate this, inspired by Kuhn's *Structure*, I will work my way through the issues to help foster the apparent revolution itself. I will demonstrate that only through a new science framework can we *scientifically* know we have built a machine that observes like we do, and when not. We currently have no idea and if we keep going the way we have, we never will.

2.1 Description(what) vs. explanation(why)

At this point I have to stop and calibrate my use of the word 'explain' in a technically specific sense. Later I will be arguing how science never actually explains anything. This sounds preposterous until one understands the technical sense of the usage of the word. If you were to work within science and observe the coal-face activity as I have, you will be amazed to know that the great bulk of scientists are essentially unaware that we only ever describe the natural world. We get implicitly trained (by imitation) to do it and we never question it in any routine or formal way. With these descriptions we become powerfully predictive of the natural world. It works. But this description cannot be confused with explanation of the natural world. Descriptions say *what* happens in the sense that it is predictive of how the natural world will appear when a scientist encounters it. This encounter is an encounter with 'phenomena', which are the natural world captured by and within the experiences of the observing scientist; the scientific observer. Explanation, in the sense meant here, is utterly different. Formal explanation is a yet-to-be defined scientific behavioural outcome that delivers an account of what necessitates that the natural world behave the way it does (in the eyes of the scientific observer that is equally necessitated by - constructed inside - that same world). This is a subtle point, and no-one puts it better than Earnest Nagel:

> No science (and certainly no physical science), so the objection runs, really answers questions as to why any event occurs, or why things are related in certain ways. Such questions could be answered only if we were able to show that the events which occur must occur and that the experimental methods of science can detect no absolute or logical necessity in the phenomena which are the ultimate subject matter of every empirical enquiry; and, even if the laws and theories of science are true, they are no more than logically contingent truths about the relations of concomitance or the sequential orders of phenomena. Accordingly, the questions which the sciences answer are questions as to how (in what manner or under what circumstances) events happen and things are related. The sciences therefore achieve what are at best only comprehensive and accurate systems of description, not of explanation.

[E. Nagel, 1961, Page 26]

The description(what)-explanation(why) divide is inconsistently understood and camouflaged by our own behaviour. To show how buried the distinction is, I will use an allegory that is floating about on the Internet. An old fish swims past two young fish and asks them, rhetorically, *"How's the water?"* When the old fish is gone, one young fish says to the other *"What's water?"* Notice in the Nagel quote, even as it is recognised that formal explanation is absent in science, there is, in the words, an implicit assumption. Consider *"...in the phenomena which are the ultimate subject matter of every empirical enquiry; and, even if the laws and theories of science are true ..."*. In these words Nagel hands us a glass of paradigm water. There is a tacit assumption that to do science is to describe phenomena. Einstein hands us $E=mc^2$, a description. We all know that the natural world shall appear consistent with $E=mc^2$. But nothing in it says *why* $E=mc^2$. The statement is a tautologous but predictive description, not an explanation in the sense meant here. Nagel's tacit presupposition is that such description defines *all* of what science does, and that you are therefore not being a scientist if you do anything else. Well, *"Says who!?"* I am here to question that assumption and ultimately refute that kind of claim about science. Science can both describe and explain, and the fact that it currently only describes is a mere accidental cultural blockage, not an immutable rule read off a rock somewhere. In science, our confinement to description is not even something as fashion-laden as an argued policy or an explicit preference or choice.

Over the coming chapters, I will show how in the scheme of revolutions in scientific structure, there can be only two. Revolution 1 got us started on descriptions - the 'what'. Revolution 2 has not officially happened, but is about to (if I read the signs correctly), and it adds 'why' by altering scientific behaviour to allow it, in addition to what we already do. What is truly bizarre (another Kuhnian anomaly) is that some of us scientists already do the new behaviour but don't realise it. In the feral wilderness of 'laws of nature' there are, currently, two species confused as one. In the proposed upgraded framework for science we will have the same natural world, with scientists embedded in it, but with two different but side-by-side and intimately enmeshed accounts produced and managed by the science-to-come; a science that has matured and moved on after millennia of development.

Chapter 3

Consciousness

In the absence of a properly established paradigmatic science framework inclusive of a widely known and trained account of consciousness, with its centrality to scientific behaviour recognised, it is necessary that I set about defining what I mean by consciousness in a technically specific way. Kuhn referred to consciousness with the terms 'the sensations', 'the given' or 'immediate experience' which are now superseded. Specific terminology will ensure that from this chapter onwards, everyone will know what I am actually referring to: P-consciousness. The intent here is to carefully deliver an appreciation of the relationship between P-consciousness and scientific observation. We can then use that understanding to examine the state of the science of consciousness. That examination does not constitute a theory of consciousness and you will not find such a theory in this book.

The cited references here are intended as a guided launching place into the ocean of technical literature on the topic. Using the referenced monograph and review works, all the details presented in this chapter can be investigated more fully.

3.1 The modern scientific perspective on consciousness

The terminology comes from philosophy, and quite recently. Its youth is an important indicator of change discussed elsewhere here. The key phrase to internalise is 'phenomenal consciousness' [Chalmers, 1996], shortened here to P-consciousness [N. Block, 1995]. P-consciousness is a technically specific term referring to the subjective, private, first-person-presented, experienced qualities of human mental life. Another common way that P-consciousness is described in the philosophical discourse is to say that it is 'like something' to be P-conscious [T. Nagel, 1974]. For example, in each case of the sensations of the redness of red, the sadness of being sad, and so forth, it is like something, in the sense meant here, from the first-person perspective. When it is like something, then there is P-consciousness.

A key element of this definition is that when the word consciousness is used, it refers to nothing more than a unified collection of different kinds of P-consciousness. Consciousness is, therefore, a whole comprised literally of parts, each of which is a P-consciousness of a particular kind.

Some people may find their ideas of consciousness challenged by the specificity of the term P-consciousness. A common confusion is between P-consciousness and 'states of consciousness'. States of consciousness are overall brain states such as coma, slow-wave sleep, REM sleep, vegetative, dissociative state and so forth. The science of states of consciousness is not the science of P-consciousness, although each informs the other. If you are in a coma, all P-consciousness is gone, and it is 'not like anything'. Not being in a coma is necessary, but insufficient, to generate any kind of P-consciousness. States of consciousness refer to the level of health, awakeness and alertness of the creature under discussion.

Another common confusion is to introduce terms like 'awareness'. Such terms have to be set aside, for they introduce a presupposed relationship between P-consciousness and knowledge/behaviour. This confusion has no place here. A human can be in a fully awake, aware state replete with all P-consciousness, and be completely unable to behave in any way. This is called 'locked in' syndrome and can result from anesthesia accidents. A similar state of affairs might also be said to occur in a child that has no knowledge and no language, yet is exquisitely connected to the world by a vivid and sophisticated P-consciousness. Terms like awareness can be ambiguous and misleading and are to be avoided. The general conclusion here is that no particular knowledge or behaviour is formally necessary prior to the existence of P-consciousness.

When reading the literature, other terminology will be found, including 'qualia' [Tye, 2009] or sometimes 'phenomenality' [Ned

Block, 2003]. The oldest term is qualia (singular quale, pronounced *'kwah-lee'*). Introduced by C.I. Lewis, he described qualia as 'recognizable qualitative characters of the given' [Lewis, 1929]. I will only use the term P-consciousness. P-consciousness contrasts with A-consciousness (A for Access), which does not refer to consciousness but rather refers to unconscious knowledge/behaviours (dormant or contributing) which co-exist with P-consciousness, but on their own have no subjective qualities [N. Block, 1995]. One can imagine this by considering that when we go into a dreamless sleep, all our memories are still present, but are not necessarily being accessed. When we dream, we have some P-consciousness that may be loosely related to memories, and our dreams may or may not be remembered. When we awake we continue to be the same person. Consider how the learnt capacity to play tennis contributes nothing to subjective life until you 'access' the 'tennis playing faculty' by imagining tennis or actually playing tennis. Then a huge raft of automatic behaviours take place without any invocation by P-consciousness, but which results in rich P-consciousness, in the tennis player, of the game. In place of A-consciousness, the term 'psychological consciousness' or even 'computational consciousness' may be found [Chalmers, 1996]. I will not be using these terms here.[a] The general conclusion here is that memory is not formally necessary prior to the existence of P-consciousness. Rare human cases exist where the hippocampus is removed, and in that case the subject can form no new memories and yet have a vivid and complex P-consciousness.

Consciousness is a unified composite of the following non-exhaustive list of individual P-consciousnesses:

(i)   Visual.
(ii)   Auditory.
(iii) Smell: olfaction (many ...).
(iv) Taste: gustation (sweet, bitter, salty ...).

---

[a] For detail, a recent review [Zeman, 2001] and a Blackwell monograph [Velmans and Schneider, 2007] are recommended. Steven Lehar produced a good example of the struggle neuroscience has had in the quest for an explanation of P-consciousness [Lehar, 2003].

(v)   Touch: (pressure, temperature …).
(vi)   Situational emotion: (mad, bad, glad, sad …).
(vii)   Primordial emotions: (hunger, thirst, fear, orgasm …).
(viii) Imagined, dreamt and pathological versions of all of the above.

If any of these are 'like anything' to say, a subject X, then X is P-conscious to some extent. For example, if in a dreamless sleep there can be none of the above and X is unconscious. If X is dreaming then experiences are being had, so X is P-conscious even though asleep. Each of these types of experience has a qualitative feel to it from a first-person perspective. Each of these can be accurately conceptualised as a perceptual 'field' or 'view' from the first person perspective. Each has its own characteristics, which can be a little challenging to encounter. For example, as a 'view', sadness is the same wherever you are (it goes with you). Technically the experience of sadness can therefore be called homogeneous. Also, as a 'view', sadness pervades your perceptual life the same way from all directions. For example, sadness does not come at you directionally, say, from under your feet. Technically this omni-directional property of sadness is called isotropy. Different P-conscious fields have different kinds and degrees of homogeneity and isotropy.

Contrast sadness with the visual perceptual 'field'. When a normally sighted person closes their eyes, their visual field may be replaced by a roughly hemispherical blackness. Before and after closing eyes there is no visual experience of what is outside that hemisphere. The brain goes to the trouble to construct a visual field only for that roughly frontal hemispherical region centred on the viewer, independent of eye direction but dependent on head orientation. When the brain does create a visual experience it can be highly inhomogeneous. It can alter radically as you move about in the world and is therefore different at different positions. Also, visual experience is radically different in different directions, reflecting anisotropy in the world around you. If you look up you may see clouds or the ceiling. Look down and you may see your feet or a chasm.

In addition, the subjective quality of the experience that is each P-conscious field is radically different in ways that are hard to describe and this is the reason for the 'like something' terminology. Sadness is radically different from redness, and both are radically different from pain or disgust. It is helpful to imagine that the brain uses a palette of such qualities to paint a picture that becomes each perceptual field. Each perceptual modality has its unique palette and a differently shaped blank canvas. We know this to be the case because in neurology there are conditions where unusual cross-modal connections can be made. The wrong pallet is used to paint a 'perceptual blank canvas'. People for whom this is the case have a condition called synesthesia and can experience, say, all number fives as blue or taste heard names. It can be profoundly perceptually enhancing or quite disturbing, depending on the particular cross-modal connections, which neuroscience has made some progress in isolating to unusual connections between particular brain regions.

> The sensation of seeing is, for us, very different from the sensation of hearing, but this cannot be due to the physical difference between light and sound. Both light and sound are, after all, translated by the respective sense organs into the same kind of nerve impulses. It is impossible to tell, from the physical attributes of a nerve impulse, whether it is conveying information about light, about sound or about smell.

> Richard Dawkins [Dawkins, 2006]

Of great significance in understanding P-consciousness is that all the different kinds of P-consciousness are differentiated from each other, yet operate as a unified (integrated) whole. Whatever P-conscious fields there are, their unity is an integrated state that is highly informative. Of all the possible ways the world could be presented, your P-consciousness has ruled out a gigantic number of them in stabilising on a particular collection at any moment in time.

A key physiological fact is that all P-conscious fields listed above are delivered by quite localised subsets of cranial central nervous system (CNS) excitable cells [Crick, 1994]. Therefore, neuroscience tells us that P-consciousness is 100% a result of the neural activity of the brain. This is counterintuitive. It feels like you perceive your hands where they are located in space. The world looks visually as if it's 'out there' and centred on your eyes. But that is mere appearance. Properties like redness are not properties of the natural world 'out there'. They are like that because your brain presents it to you that way.

We have all experienced back pain called 'referred pain' that appears to come from a position other than the region of original insult. The brain and body physiology is not perfectly equipped to infallibly locate and accurately project experiences 'as if' they come from the correct location. This reality of brain operation does not entail a denial that the external world is involved in the final presentation of P-consciousness. Of course it is involved. All that is claimed is that the (imperfect) process that makes the external world *like something* is contained within the skull.

A useful way to imagine this rather mysterious process is to pretend that your brain is a lens. You are not looking through a lens at something. Instead you *are* a lens that is being constructed, moment to moment, with the image already in it. It is an image that is only visible from the perspective of *being the lens*. One lens image per P-conscious field. If you can 'get your head around that', so to speak, then you'll make sense of the task at hand.

P-consciousness can misrepresent, be damaged and can be disrupted by drugs or illness (consider hallucination as a distorted P-conscious visual or auditory field). The measurements made at your retina are not what you visually experience, which is actually a construct of the visual (occipital) cortex at the back of the brain that projects the experience as if it were centred on your eyes. Similar localisations of all perceptual fields put the neurological origins of the experience at different positions in the cranial brain material, and all of the perceptual fields (experiences) are re-positioned (mapped) to variously appear centred on the head or on the position of the sensory apparatus (e.g. the ear drums or the finger tips).

To reinforce the reality of the localisation of P-consciousness to the brain, consider 'phantom limb syndrome' (P-consciousness depicting nonexistent body parts) and 'blindsight' (successful visually guided manipulation of body parts without any visual P-consciousness) [Velmans and Schneider, 2007; Zeman, 2001]. In both these cases the P-consciousness and the originating measurements (sense organ signalling) are dysfunctionally related in a way that reveals the separation of the sense organ and the P-conscious field normally associated with it. Localisation of P-consciousness origination in cranial CNS means that P-consciousness is *not* delivered by spinal CNS or by the peripheral nervous system(s), which includes the huge nervous system lining the gut. This is decades (in some cases century) old physiology. The primordial emotions are those associated with the ancient basal brain regions involved in an essential bodily regulatory process called homeostasis. P-consciousness fields for homeostasis include hunger, breathlessness, thirst and so on [Denton, 2005]. These contrast with the situational emotions of sadness, happiness, anger, disgust, fear and so forth, which have at least some localisation outside the ancient basal brain.

Consider a perceptual modality that is a particular kind of P-consciousness. For example auditory P-consciousness. It is common to encounter, in the nascent science of consciousness literature, the term 'contents of P-consciousness' or simply 'contents of consciousness'. If you hear a bell, then you are experiencing the sound of a bell, and that particular subset of all sounds is one of the 'contents of auditory consciousness'. Contents of consciousness are literally those specific subsets of the modality that the subject attends to or otherwise reports as having being experienced. A chimpanzee may press a button indicating that a yellow blob has been visually perceived. At that moment the contents of the animal's visual P-consciousness was identified, by the animal, as containing what its brain has labelled with what we humans identify as yellowness.

Notice the deeply problematic nature of the identification of a subjective quality. All that can be said, as a result of receiving a report of yellowness, is that the subject has had an experience reported as yellow, which may or may not be 'like' another subject's experience with an identical sensory impact (say on the retina). In each subject's life, they have learned to report particular experiences as yellow, and there is no way (in the absence of a future suitably sophisticated science of consciousness) to know how similar or different each subject's experiences are. Indeed there is a large body of evidence that, for example, particular qualities like colour are actually created by the need to distinguish and report (behave in respect of) colour. For example, a particular ethnic group might have 13 words for green or 15 words for white, but neither group can perceive or report blue as a distinct colour, and they have no word for it. The encoding of the environment into colours is highly culturally specific. It seems to be turning out that musical pitch, timbre, consonance and dissonance have a similar acculturated allocation of subjective qualities.

P-conscious fields have a non-unique relationship with the sensory measurements used to generate them. From the point of view of understanding how this impacts an attempt to use a scientific account of P-consciousness as a scientific account of a scientific observer, consider two extremes. The first is the gestalt 'duck-rabbit' illusion (see Figure 3.1). You, the observer, stare at a picture and have a constant retinal impact resulting from the one drawing. Yet you can literally have two experiences. One of a duck. One of a rabbit. The inverse is also true. It is possible to wear glasses that invert your normal retinal impact. When you first put these glasses on it is a very disturbing experience. Everything is upside down. If you persist, however, eventually your

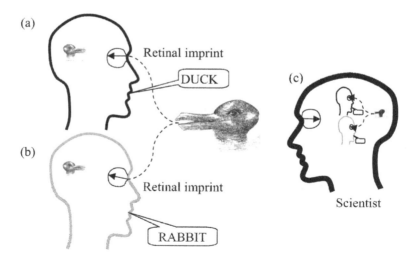

Figure 3.1.   The scientific study of consciousness. Two subjects (a) and (b) have identical retinal imprints from the one stimulus and then report seeing two different things. The scientist (c), usually excluded from these diagrams, is doing the science of consciousness (as it is currently defined). The scientist observes (a) and (b) behaviour, and has no access to the P-consciousness of the subjects. Scientist (c) only has P-consciousness of the verbal reports and of the stimulus.

brain inverts the image and your visual P-consciousness is normalised. In that state you have a different retinal imprint, and end up with the original P-consciousness. Thomas Kuhn was well aware of this, and it figured in his account of the role of gestalt changes in the world view of scientists [Kuhn and Hacking, 2012, Chapter X].

Figure 3.1 shows the relations between the external world, sensory impact, subjectively experienced qualities, resultant behaviour, and their relationship with scientific observation. It uses the duck-rabbit gestalt illusion to ensure we recognise that the retinal impact and the mental experience of vision are not uniquely related. In Figure 3.1 there are two scientists, (a) and (b), acting as experimental subjects. They are observing and reporting 'what it is like' to another scientist, Figure 3.1(c), a scientist that is usually absent in such diagrams. Scientist (c)'s observation is also shown as contents of (c)'s P-consciousness. (c) has no access to the P-consciousness of the subjects. Instead, only the verbal reports are accepted as evidence of the subjects undergoing the gestalt

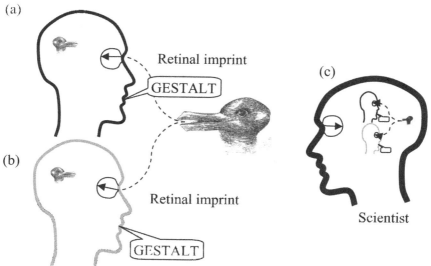

Figure 3.2. The scientific study of scientific behaviour. Two scientists (a) and (b) study the illusion, have objectified the phenomenon and named it gestalt. Both report the same thing: that a gestalt illusion is being had. They are doing science on Gestalt illusions too. They have done this by acting as each other's independent observer. The result is identical to Figure 3.1 except that Scientist (c) is now not studying consciousness, but is actually studying scientific behaviour.

illusion. It is (c) that is doing science, it is (c) that is doing the scientific observation, and the science is the science of consciousness.

The same physical situation is shown in Figure 3.2. Two scientists (a) and (b), recognising the illusion, have objectified the phenomenon. Both report the same thing: that a gestalt illusion is being had. They are doing science on Gestalt illusions too, and they have done this by each acting as the other's independent observer. The result is similar, but not quite the same as what happened in Figure 3.1. Where does this leave Figure 3.2(c)? Where scientist Figure 3.1(c) was studying consciousness, scientist Figure 3.2(c) is actually studying scientific behaviour.

The purpose here is to familiarise the reader with the kinds of perspective shifts needed to understand the science of consciousness, objectivity, the science of scientific behaviour, and the role of P-consciousness in scientific observation.

3.2   Objectivity and the first-person perspective (1PP)

Objectivity helps us select and agree on, name and categorise perceived entities as the target of a scientific study. We scientists *implicitly* understand the role of objectivity as a kind of 1PP suppressant. This process is deeply ingrained by being told when it goes wrong and by experiencing the effects of its failure. Objectivity is a behavioural norm operating without a scientifically established account of what it is physically doing for us. A classic method of objectivity is the 'double-blind' experiment where neither test subject nor scientist has awareness of some crucial detail in the mechanism being studied. We scientists are trained in such things as 'selection bias' and complicated statistical methods that reveal the actual weight of evidence, where less rigorous analysis of data might see wishful thinking seeping into the results. Procedural techniques abound, but ultimately, a deeply ingrained sense of objectivity is operating in scientists in the same way that we don't have to understand how a car works to operate a car, and we don't have to think much about it while we are doing it.

In a future where a scientific account of a first-person perspective (1PP) is normal and technically nuanced, it will be routine to expect all scientists to have been trained in the role of one's own 1PP in an ability to do science. We will not only understand that objectivity works, we will understand how it works. The technically understood centrality of a 1PP in the life of a scientist will change our understanding of objectivity without replacing it or downplaying it. We can take a look at that future without having a full scientific account of the 1PP. We can do this by simply allowing ourselves to recognise the 1PP as a real physical process delivered as P-consciousness by means unknown, and that the resultant 1PP is all we have (scientist and non-scientist) to access the natural world.

To see how subtle this process is, consider a sentence like *"Our group is interested in natural phenomenon Thing-X"*. That very sentence, uttered a thousand times a day in science, has already broached and bypassed the 1PP. Actually, that's probably wrong. The words 'natural phenomenon' will likely not be there, leaving *"Our group is interested in thing-X"*. That has eliminated the role of *phenomenal* consciousness from the process. 'Phenomena' are contents of P-consciousness. 'Thing-X' is out there in the world, but we only know that because it (or its causal descendant) has become contents of our P-consciousness. It is contents of P-consciousness that is being referenced, and without P-consciousness there would be nothing ... as in 'no thing' ... to discuss. That is how deeply embedded this issue is. Let me rewrite that sentence with all the lost details added. It comes out like this: *"Each scientist in our group has examined our 1PP, P-consciousness, and mutually agreed (via a discussion mediated by the P-consciousness of each of us) that there exists a natural process operating outside all of us, that is suitable for study and that we have labelled 'Thing-X'. We set about creating statements that are predictive of how Thing-X will appear generally in any 1PP (P-consciousness) that cares to encounter it."*

This is how we render the 1PP and P-consciousness invisible in our daily lives. Objectivity thus imbues our every action and is in place before we even open our mouths. And for the entire history of science that brevity and conciseness has served us well and will go on as such. However, how is a science of the 1PP, a science of P-consciousness, that explains how we can even begin to operate as scientists, be done when we can't even utter a sentence that includes it in anything we do?

Let's go back to our group of scientists intent on studying Thing-X. Let's fill a lecture theatre with 400 awake, alert scientists that study it. On a table at centre stage, is Thing-X. In that lecture theatre, how many objective views (N_o) and subjective views (N_s) of Thing-X are there? If I was asked, I would say $N_o = 0$ and $N_s = 400$. How about you? Every scientist in that theatre has agreed there is a 'Thing-X' and if pressed, each would say that they are taking an 'objective' view of Thing-X. But that objective view does not physically exist. It is an abstract agreement by the group; a shared concept, all agreed to, that identifies Thing-X within the 1PP of each scientist. So what is this objective view? It is actually an abstract collective view from nowhere. No individual scientist has such a view. It doesn't mean objectivity is nonexistent. It merely means that the objective 3rd-person view, the 3PP, does not actually exist anywhere. Objectivity is a behavioural norm and does not involve any sort of actual view. It is a 'virtual' view. It is *as-if* there is such a thing, but in fact there is no such view being had. There is only a 1PP and each of us has one and in each of us it is unique. In that lecture theatre there are 400 *different* 1PP views of Thing-X, and it is impossible in principle for that number to be less than 400. Each scientist is in a different seat, and the view is from a unique position. They may be very similar, but none are the same.

This leads to the next nuance: a perspective is not necessarily a 'view'. A 'view' is an experienced perspective taken of something from a particular position. In the case of 1PP we mean *'from the point of view of being'*. In the case of our 400 awake alert scientists, each has such a perspective view, from the point of view of being a scientist seated in a lecture theatre. Down at the front of the lecture theatre (in your mind's eye, a perspective view that I am creating in your P-consciousness), Thing-X sits. What about the 1PP of Thing-X? There is a perspective from the point of view of being, but is there an actual view? Remember, I am thinking of a view as any P-consciousness whatever. For example, what if the only P-consciousness of Thing-X was 'sadness'. That is what I mean by a view from the perspective of being Thing-X.

If Thing-X was a dog, then one might argue there is a Thing-X 1PP 'view'. If Thing-X was a rock then one might argue there is no Thing-X 1PP view. There is a tendency here to confuse the existence of a perspective with the existence of a view from that perspective. Be it rock or dog, Thing-X is located in the world and that location imbues Thing-X with a perspective of everything that is NOT Thing-X. The perspective is innate and unavoidable. It is the perspective acquired from the position of Thing-X, from *being* Thing-X. But the mere existence of a perspective cannot be confused with, nor does it justify the existence of a *view from that perspective*. The fact we can imagine a view from the point of view of being Thing-X – the '1PP of Thing-X' – is irrelevant in this respect.

If we have a science of consciousness, then what it is trying to do is account for whether or not, due to the organisation of Thing-X, there is a *view* from the point of view of being Thing-X. We can now understand better the goal that is the responsibility of a science of consciousness: it is to account for how and why a portion of the natural world may have a view, or otherwise, from the innate perspective of *being* that portion of the natural world. How is that science going?

3.3 The science of consciousness

Clearly the empirical science of consciousness has done well in delivering the picture of the previous paragraphs. But as the science matures, and the investigation tracks ever deeper into the fine structure of the brain, a systemic problem has begun to exert control. To see it, we first need to be aware of a half century old philosophical position called the 'Mind-Brain Identity Theorem' (MBIT), which is roughly that 'to describe the brain is literally identical to describing the mind' [Borst, 1973; Chalmers, 2000; Churchland, 1988; Feigl, 1958; Macdonald, 1989; Smart, 2004]. This is not a statement of proved natural regularity. It is a statement of opinion. Citing it in a science of consciousness context results in a kind of 'get out of jail free' card, allowing what you've always done to account for something you've never had to account for as a scientist: a first-person perspective/P-consciousness/the scientific observer. The unusual nature of this proposition is that at no stage has anyone felt the need to define a '*what-it-is-like-to-be-a-rock/rock identity theorem*', where presumably to scientifically describe a rock is equally prescriptive of what it is like to be a rock. A fully explanatory account of consciousness should be able to formally state the 1PP of a rock (including the lack of a 1PP), just as well as it accounts for the 1PP of a dog or a human. A rock or dog's inability to report is moot to the argument. The oddity of this situation is there to be seen. This is just the tip of an iceberg of misdirection caused by philosophy delving beyond its borders into science and we scientists letting them determine what is going on, rather than studying ourselves. This is the place inhabited by the science of consciousness.

In the science of consciousness as it is practiced to date, two observed phenomena, (i) ABC and (ii) a report of an experience, are being correlated. ABC might be a neuron firing. The report might be 'red'. Two objectified phenomena are correlated in the usual way. It sounds reasonable, yet the only way this can ever be held as an account of a first person perspective is to presuppose that the MBIT is a 'law of nature', whether even aware of it or not. In this dodgy shadow of an evidence boundary condition, the science of consciousness deflects criticism by re-branding itself as 'The ABC-correlates of consciousness'. If all you are doing is declaring you've found a physical mechanism that correlates with a report of P-consciousness, then you can remain mute on the absence of a real account of P-consciousness and be predictive in a way that resembles what you've always done. By carefully attending to the details of the report and the measurement, all sorts of subtle phenomena become evident. That is what has been going on for over 20 years. Science has done its best to make it look like what the centuries-old paradigm says 'normal' science looks like. But it's on shaky ground. We can all be happy that science has even done this much when in the past the more likely outcome was to be marched out of the lab for suggesting that you do anything with consciousness. We have made progress. The science is interesting. It has taught us a lot. It continues to be valuable work. But in doing so we have crossed a line, and it's a very important line. An under-examined line.

The key to the anomalous status of the science is that the activity has a fundamentally different goal: an explanation of something that science has never attempted to explain, ever: a first person perspective. Furthermore, it's the exact same thing upon which 100% of science is critically dependent for scientific observation: P-consciousness. To see the difference between what has gone before and what the science of consciousness is doing, consider a classic: Newton's second law *F=mA*. Two objectified observable *phenomena*, force F and acceleration A are examined. Voila, the 'force correlate of acceleration' emerges and is written down. The rest is history. But take another look. There is something missing from the science. Nobody thought to ask *"What is it like to be the mass m?"* Nobody thought that such a question is within the auspices of valid science.

In the science of consciousness, in effect, science is accepting F or A and their correlation as an account or explanation of what it is like to be *m*. To see this, let's revisit the 'ABC-correlates' paradigm. We measure ABC. We measure a report of an experience X. We get the 'ABC-correlate of an X report', which is a result of asking the test subject 'what it is like to be the thing being tested'. We avoid the glaring inconsistency between the evidence and what is being explained by using the correlate terminology. It is exactly like asking *m* to tell us what it is like to participate in *F=mA* when the only difference is that *m* cannot report it and no-one ever thought it a valid question. It doesn't mean there is nothing to report. Indeed the report X = *"It's not like anything"* is a valid outcome for *m*. The strangest thing of all about this is that somehow a relationship of the kind *F=mA* is expected to account for a 'what it is like' without any principled argument that the process speaks to it, and when we have centuries of never expecting there is one. Instead, there is the MBIT floating about: no scientifically proved principle whatever. What if the *m* is you, an astronaut after blast-off, experiencing a 10G acceleration. Would *F=mA* be expected to account for how you felt during that process? If not then how is the *'neural correlate of report of feeling a 10G acceleration'* any more explanatory of how it feels?

Now let's look at the long term outlook for ever more sophisticated 'ABC-correlates' science. Let's say we get to the point where we are literally able to say something along the lines of *"When this, and only this exact cohort of neurons dance about <exactly like this>, then we get a red report, and when the identical neurons <dance like that> we get a green report"*. We can arbitrarily control the outcome from green to red and back. It's repeatable and predictable in a way that would make any empiricist proud. We have lots of identical test subjects, reams of data, lovely graphs, we are statistically rock solid. And we are sitting there, completely devoid of all explanation of why it is 'like that' for the subject. If an engineer wanted to build a machine with a red/green experience out of silicon, you would have no clue what to tell them. Your science has delivered exactly what you asked for: description, not explanation, and the thing that is to be explained is unlike anything else in the history of science. Your cherished method doesn't touch it.

This is the method, and the state, and the eventual fate of the 'ABC-correlates' paradigm. It is science in the sense that it does what science has always done. But it's not any sort of explanation of consciousness. Instead, it has revealed the true nature of the science paradigm as a describer, not an explainer. If science was to actually explain the first-person perspective, and it was successful, what would the scientists be doing? The answer is *"Not what we are doing in the previous paragraphs. It must be something different"*. Is the suggested alternative science possible? Well one of the few things we can say for certain is that it will be impossible if you never let yourself ask and answer the question *"Is the suggested alternative science possible?"* This book is about answering that question.

> Any new interpretation of nature, whether a discovery or a theory, emerges first in the mind of one or a few individuals. It is they who first learn to see science and the world differently, and their ability to make the transition is facilitated by two circumstances that are not common to most other members of their profession. Invariably their attention has been intensely concentrated upon the crisis-provoking problems; usually, in addition, they are men so young or so new to the crisis-ridden field that practice has committed them less deeply than most of their contemporaries to the world view and rules determined by the old paradigm. How are they able, what must they do, to convert the entire profession or the relevant professional subgroup to their way of seeing science and the world? What causes the group to abandon one tradition of normal research in favor of another?
>
> Thomas Kuhn [Kuhn and Hacking, 2012, Page 143]

3.4 Other consciousness aspects

3.4.1 *Unconscious processing, skills and novelty*

P-consciousness grounds all new learning, and in particular higher-order skills (abstract reasoning and symbol processing), in P-consciousness. In the process of mastery, the skill becomes stored in non-conscious brain processes, making the trainee less reliant on P-consciousness. This is the repository for skill at tennis, for example. We all experienced this process when learning to drive. Once we acquire the skill, we are hardly aware of it except when novel events arise. In its mastery, the skill becomes largely unconscious. This leads us to consideration of whether new skills can be unconsciously learned or whether an existing skill can be enacted entirely unconsciously. There is ambiguous/patchy evidence for this in the literature, and those that claim it tend to find it a marginal effect, and replication of the results can be problematic. It remains an active area of research.

A similar effect is unconscious priming. It is sometimes claimed that, for example, unconscious visual percepts can bias our emotional state, and thereby affect future behaviour (such as a decision to buy something). Like the unconscious processes mentioned above, clear reproducible evidence remains a work in progress. The unconscious percept, the decision to act, and the link between these two things presents a suite of investigative challenges for which clarity is yet to emerge. That clarity will have no bearing on the major aims of this book and needs no further elaboration.

3.4.2 *Attention*

An ability to focus on a preferred subset of P-consciousness is to select from William James' *"bloomin', buzzin' confusion"* [James, 1890], by suppression of those that are unwanted. This is the function of the 'attention system' in cognition. There is a conscious and an unconscious component to this selection and a skill level in its usage. For example, we attend to someone talking and the rest of the room (our periphery) is suppressed to some extent. While this is going on we continue to have no visual P-consciousness of what is behind us. While the attention system is a major area of research, and is of great significance in consciousness studies generally, its existence also has no great bearing on the major goals of this book.

3.4.3 *Higher order consciousness*

There have been quite a number of journal articles and books tackling something called 'meta-consciousness', 'higher order thought', 'higher order cognition' or 'higher order consciousness'. The discussion of such things as 'self', 'knowing that we know' and being 'consciousness that we are conscious' and so forth, are considered as part of 'higher-order' processing. For example, a 'red' experience can be called a 1st-order process. An ability to report red linguistically might be a 2nd order process.

For the purposes of this book, higher-order processing has no immediate usefulness. In the absence of any clear evidence for unconscious learning, we must accept that higher order metacognitive processing cannot have arisen from nothing. Whatever you might think higher-order processing is, it is ultimately acquired through grounding in P-consciousness, like all other learning. In the grossest sense, take P-consciousness away and *all* thought and *all* consciousness is gone, including 'higher-order' versions of them.

The fact that we have higher-order consciousness facilitates our ability to ask questions like those being asked here. However, an account of the basics of P-consciousness does not need higher-order ideas. Higher-order processing is not a matter of kind. It is merely a matter of scale or degree. Whatever we have as P-consciousness, it is likely that most vertebrates have some variant of the same thing, albeit in a less complex, perhaps dimmer way. An ultimate theory of intelligence will account for the way P-consciousness arises, and *then* how it may be configured for increasing levels of meta-awareness such as the self and knowing that you know. Higher-order ideas, if encountered, cannot be confused with a scientific account of P-consciousness. These concepts will have no bearing on the outcomes of this book and are set aside.

3.4.4 *The 'self'*

In consciousness studies, 'self' is a particularly massively-analysed concept, and its literature is tortuous and confusing. I can offer relief: *forget about it.* 'Self' cannot be classed as causal of or necessary for P-consciousness. Indeed, it is obvious that the reverse applies. If you have any sort of rudimentary P-consciousness, then a 'virtual or as-if self' is immediately created by projection. All P-consciousness is a 'view delivered from the point of view of being' the entity doing the generating. That 'view' goes with the entity, wherever it goes. Take all P-consciousness away and the self is gone. Acquire knowledge using P-consciousness, and that knowledge contributes to the level of sophistication of the self (expressed in sophistication of behaviours directed at the self). By concentrating on an account of P-consciousness, the concept of self is automatically addressed. Talking about self before sorting out P-consciousness is interesting but premature. It will be set aside for purposes here.

3.4.5 *Primary consciousness*

Where the term 'primary consciousness' is encountered, it can be taken to mean P-consciousness. There is nothing over and above P-consciousness that primary consciousness can refer to. 'Primary' consciousness concepts will not appear again here.

3.5 Summary

In this chapter we learned that the technically specific term for consciousness is P-consciousness and refers to the experienced, first person perspective (1PP) that constitutes our mental life. If you are dreaming then you are P-conscious yet asleep. P-consciousness is a construct of the brain, not the eyes or the ears or other sensory apparatus or the spinal column. It was recognised that P-consciousness is an identity with 'scientific observation' in the scientist. This does not mean that all P-consciousness is scientific observation! The pain of accidentally banging your thumb with a hammer during carpentry is certainly a subset of your P-consciousness, but it is not an act of scientific observation unless associated with the particular behaviours surrounding a scientific act. All scientific observations are an identity with P-consciousness. Scientific observation is a behaviour-specific subset of all P-consciousness.

While objectivity operates to abstract 'contents of P-consciousness' into an observer-independent form, at no time does an actual 'objective view' exist. Objectivity serves to 'objectify things out of a scientist's subjectivity' for the purposes of observer-independence of the resultant scientific descriptions.

We learned that it is the job of a 'science of consciousness' to create a scientific account of P-consciousness, and that the science is presently conducted under the generic banner 'ABC-Correlates of Consciousness'. The science is therefore trying to account for, with scientific observations, an ability to scientifically observe. This is a logically flawed expectation. It also locates the science of consciousness at a scientific evidence boundary condition unique in science. Unlike anywhere in science, the same scientific behaviour has an extra demand placed on it: a demand for an account of the 1PP of the test subject that, for the same scientific behaviour elsewhere in the natural world, is not required.

In this way we can now see that the ABC-correlate science of consciousness is actually a disguised, logically compromised science of scientific behaviour (the scientific observer) operating at an evidence boundary condition where it is trying to account for something extra that no other science context demands of it.

Therefore, when we examine closely our relationship with objectivity, our beliefs about the knowledge obtained by the use of it are deeply paradoxical. We hold objectivity in high regard and the resultant successes lead us to think there is an objective view when there is none. All along, objectivity is actually objectifying something out of subjectivity and yet we methodologically deny there is evidence for subjectivity, by looking for evidence outside ourselves, where we usually find evidence. Yet the act of objectivity itself literally evidences the subjectivity at the heart of it. We scientists, literally, are the scientific evidence for P-consciousness, and yet we are so intent on placing evidence origins outside ourselves, that it never occurs to us that we scientists are part of the natural world too, and must be evidence of something, yet that evidence is never seen for what it is.

This unusual state of affairs is accepted here as scientific evidence that scientific behaviour itself is anomalously configured and in need of attention. This is an opportunity to question the deeper nature of scientific behaviour itself.

> We have already seen, however, that one of the things a scientific community acquires with a paradigm is a criterion for choosing problems that, while the paradigm is taken for granted, can be assumed to have solutions. To a great extent these are the only problems the community will admit as scientific or encourage its members to undertake. Other problems, including many that had previously been standard, are rejected as metaphysical, as the concern of another discipline, or sometimes as just too problematic to be worth the time.

> [Kuhn and Hacking, 2012, Page 37]

Chapter 4

The Route to Normal Science

In 'Structure' Chapter II, Kuhn reveals a pattern, since antiquity, of repeated instances of a field of study coalescing from pre-paradigmatic philosophical beginnings to embark on its scientific history proper. The pattern is shown diagrammatically in Figure 4.1. History proper, the record reveals, becomes a series of paradigms punctuated by anomaly and crisis resolved to a different successor paradigm. And so forth. Transitional periods may appear gradual or relatively rapid depending on your position inside or outside, and over what time period observations are taken.

Furthermore, the early pre-paradigmatic era is typically a transition

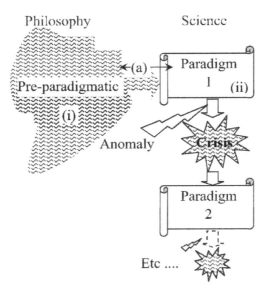

Figure 4.1.   The dynamics of creation and evolution of a science from philosophical roots.

from a sub-topic within philosophy to a science with specific concerns and a coherent tradition. The classic exemplar is the 'dynamics' studied in antiquity by multiple scholars that then became the paradigm of Newtonian dynamics in the 17th century, which underwent a 19th century crisis from which emerged a new paradigm of 20th century relativistic dynamics. Each era evolves its practice, concepts, known phenomena, known unsolved problems and techniques from which current practitioners draw guidance.

Within pre-paradigmatic antiquity, there is no 'normal' science. Instead, there is a set of competing schools of thought. When a particular school attracts enough followers, the chosen view dominates and a science paradigm is born. Its practitioners have adopted what is termed 'normal' science; the science of the paradigm. Pre-paradigmatic eras have no standardised methods or ability to even recognise natural phenomena as within the purview of the nascent discipline. The early dialogue is recognisable through its tendency to reiterate fundamentals with each new exposition, and to build the field anew from its foundations. The earlier the era, the less recognisable its output as science, and the more prevalent is the tendency for the writings to be addressed at other writers, rather than the natural world that is actually the underlying topic. Multiple schools of thought thereby battle for dominance of the era, with nature somewhat sidelined in the confusion.

Now consider the science of consciousness, which would have been regarded as an oxymoronic statement at the time of Kuhn's *Structure*. The Kuhnian depiction of the pre-paradigmatic state of the science of consciousness is striking, and it is literally happening now.

The emergence of a science of consciousness can be argued to be roughly 20 years old, and this book is yet another in a recent blizzard of foundational books on consciousness. Another characteristic of the proximity of a new paradigm is the publication of specialised journals, the formation of specialised societies (and their conferences), and the appearance of the specialisation as a topic in a curriculum in a teaching context. In the science of consciousness, in the last 20 years we now have the 'Towards a Science of Consciousness', 'Quantum Mind' and 'Association for the Scientific Study of Consciousness'' conferences. We have the 'Journal of Consciousness Studies', the journal 'Consciousness and Cognition' and the 'International Journal of Machine Consciousness'. All unique, new products of the last 20 years or so, and all directed at a nascent science for which the direct association with its philosophical roots remains in place, yet diminishing in relevance, just as Kuhn suggested.

Kuhn also sees a message in the role of the book in the identification of pre-paradigmatic sciences. As work proceeds, the historically multiple schools' erroneous theories are discarded and there is a convergence on the emerging dominant theory. The community members' output begins to address nature more and each other less. As the paradigm normalises, and the practices proliferate, become more complex, and their accessibility to the lay person recedes, the book diminishes as a medium of information exchange. All these things can be said of the science of consciousness. It is probable that only the timescales are debatable. It is not always easy to see when you are inside it.

> Today in the sciences, books are usually either texts or retrospective reflections upon one aspect or other of the scientific life. The scientist who writes one is more likely to find his professional reputation impaired than enhanced.

> [Kuhn and Hacking, 2012, Page 20]

Is Kuhn's message for this book grim or not? Continuing ...

> Only in earlier, pre-paradigm, stages of development of the various sciences did the book ordinarily possess the same relation to professional achievement that it still retains in other creative fields.

> [Kuhn and Hacking, 2012, Page 20]

The message is that this book's role, if it has one, is as a means of idea transaction in a pre-paradigmatic setting, but less so otherwise. There is a recent confound to Kuhn's messages: in 1962 there was no concept of the incredible proliferation and impact of globalised information technology and online paperless publishing. We are in new territory that has yet to stabilise. The role of the book in science is in a state of flux, but may yet continue the existing pattern. Time will tell.

The science of consciousness, being probably of an age somewhere between twenty years and the age of Kuhn's book, can be seen as one in a state of flux, but with all the characteristics of his *Structure*, Chapter II. There can be seen a transition from a multitude of philosophical camps to specific scientific scrutiny outside philosophy (usually under the title 'ABC Correlates of Consciousness' detailed in the previous chapter). There is a 20 years old tsunami of books competing for some kind of theoretical primacy that is now giving way to a crescendo of technical publications that are increasingly difficult for the lay person to internalise. There are, unlike any time in history, tertiary teaching and research institutions that have consciousness on the curriculum. Everything about this can be seen in Figure 4.1 and its associated Kuhnian characteristics. To hazard a guess, I would judge us to be at Figure 4.1 position (a) for the science of consciousness.

I now draw your attention to what is missing from all this and Kuhn's prescriptions. Where is science itself in terms of Figure 4.1? This gets zero attention in Kuhn's book. We now have 'science studies' as a teaching topic and an area of research. It started around Kuhn's time. Ian Hacking even attributed its birth to Kuhn's '*Structure*' [Kuhn and Hacking, 2012]. What does it do? Consider a 'science studies' book such as David Hess's "*Science studies : an advanced introduction*" [Hess, 1997]. The summary in my local library catalogue is:

> Science Studies is the first comprehensive survey of the field, combining a concise overview of key concepts with an original and integrated framework. In the process of bringing disparate fields together under one tent, Hess realizes the full promise of science studies, long uncomfortably squeezed into traditional disciplines. He provides a clear discussion of the issues and misunderstandings that have arisen in these interdisciplinary conversations. His survey is up to date and includes recent developments in philosophy, sociology, anthropology, history, cultural studies, and feminist studies. By moving from the discipline-bound blinders of a sociology, history, philosophy, or anthropology of science to a transdisciplinary field, science studies, Hess believes, will provide crucial conceptual tools for public discussions about the role of science and technology in a democratic society.

'Science studies' is not about scientific behaviour. Science studies is about the products of scientific behaviour, assuming that scientific behaviour is a completed behaviour. Consider Figure 4.2, which looks at a scientist (b) studying scientific behaviour (a). Notice the depiction of the scientists embedded in the environment. Consider the idea that each human is embedded in the environment in the same inseparable way a human-shaped body of water is embedded in a lake of water. I need the reader to encounter this notion for reasons that will become clear much later.

Now consider the basic scientific behaviour of Figure 4.2(a) *Scientist I*. *Scientist I* observes the natural world using P-consciousness, then makes a statement (an utterance) that purports to capture an underlying regularity in nature (in this case Newton's second law) that might be conveyed to another scientist, who may then become predictive of the natural world in another setting. The viability of the law of nature critically depends on this process as a form of argumentation about the accuracy and context of the statement. At this level, understanding the neuro/cognitive biology that accounts for the ability *Scientist I* uses to

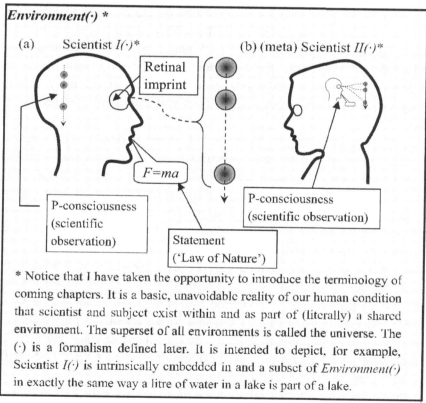

Figure 4.2.   (a) The creation of a law of nature by scientific behaviour and (b) the meta-science of it.

create the utterance is irrelevant. That the utterance happens is all that matters. The utterance may be completely wrong (mostly they are wrong) and it changes nothing.

In Figure 4.2(a), does a modern depiction of a scientist's capacities and behaviour make scientific behaviour itself a member of the pre-paradigmatic process of Figure 4.1(i) or paradigmatic as per Figure 4.1(ii)? The Kuhnian signs of it being in Figure 4.1(i) are that the exploration and formulation of the behaviour is in some way the purview of philosophers, and multiple competing camps vie for supremacy over what constitutes scientific behaviour.

While all this debate is going on in philosophy, what is going on in science? If you look, you will find scientists barely aware that our behaviour is hotly debated and none of that debate changes anything we do. No scientist need defer to philosophers. Entire careers can be undertaken in a complete absence of all input from or even awareness of philosophy. Furthermore, the behaviour depicted in Figure 4.2(a) is ubiquitous to the point that if you find any scientist not doing it or not acting as part of a team that does it, then you can claim they are not a scientist.

A theoretician, for example, might utter a fabulously complex theorem. But until that utterance is found to cohere in some way with the natural world and the logical consequences of it are predicted and then are experienced in a Figure 4.2(a) context of observation, that utterance is not a law of nature, and the utterer has not been part of a fully expressed instance of scientific behaviour. Einstein's relativity went through exactly this process in the early 20th century. How can a pre-paradigmatic philosophy debate rage about a natural world behaviour (scientific behaviour) that is clearly operating like a solidly locked in, well travelled paradigm? I conclude that science is actually in a well established paradigm and that I can just forget about the philosophical side. The ongoing pre-paradigmatic debate in philosophy must be a symptom of some other aspect of the situation of science that I hope will become clear over the upcoming chapters.

The automatic, implicit compliance by scientists with Figure 4.2(a) behaviour throughout all sciences without exception, nonstop for well over a century, indicates that the behaviour is part of an established paradigm as shown in Figure 4.1(ii). Logically, then, it must have come from a pre-paradigmatic era Figure 4.1(i). When was that? I'm not enough of a historian to know exactly, but our scientific-behaviour paradigm probably originated in the ancient Greeks and coalesced, leaving philosophy behind, around the time of the 18th century. Someone else can sort this out. It does not alter the fact that 21st century scientific behaviour itself is within a stable paradigm.

Now consider Figure 4.2(b), where another scientist has encountered an instance of undocumented (as already revealed here) regularity in the natural world: *the scientific behaviour of Figure 4.2(a)*. The behaviour is made possible by the (neuro-) biology of cognition, including P-consciousness. It includes use of as yet undocumented procedural norms and results in an ability to make specific kinds of utterances that capture regularity in the world inhabited by the scientist. The scientist is embedded in the natural world, describing it from within. This all sounds straightforward. So why is it that the natural regularity that is Figure 4.2(a) remains undocumented yet universally enacted? What if there was something implicit and limiting in the lack of managed attention to what constitutes scientific behaviour? It has already been shown how the behaviour is passed on, unmediated by any formal documentation (through novice imitation of mentors). Science seems to operate well in the absence of an explicit 'law of nature' applicable to it. What might we be missing?

4.1 An aside: The philosophy => science transition in the science of consciousness

The obvious practical irrelevance of philosophy-of-science to science generally (in the above discussion) is revealing of a functional separation of some kind. What about the presence of philosophy in the science of consciousness? In the Figure 4.1 position (a) transition, which is the claimed condition of the science of consciousness, and as an engineer intent on building artificial (inorganic) brain tissue for AGI and implant purposes, I have found myself encountering outwardly scientific books and journal papers investing much energy in a discourse of philosophical categories such as XYZism. As we transit out of that era it is worthwhile writing down what it is like to experience the process. The contention is that the philosophical discourse is becoming (or has already become) a liability that is getting in the way of science. Ordinarily I wouldn't be concerned at the involvement of philosophical discourse. It can imbue otherwise mundane topics with nuance, and my involvement in it is totally optional. It has no practical impact. Indeed that lack of practical impact is central to my objections.

One of the factors is the level of philosophical involvement in the science. In the science of consciousness, the philosophical argument about XYZisms can be found being systematically mistaken for science. It is as if XYZism discourse itself has taken on a role as a source of scientific knowledge in the area. For example, I continually encounter 'computationalism' being accepted as some kind of 'law of nature' when it is demonstrably not. This logical failure directs $millions at AGI projects that can be argued to be failed a-priori. There are other 'isms' that direct funding at projects that are arguably not science either (mostly in cognitive science and psychology). My understanding is that philosophy is, to science, a source of disciplined argument about potential knowledge. Somehow, knowledge of philosophical XYZisms, and their argumentation, despite the state of transition identified above, has become or remains accepted as in some way informing a scientific understanding of the natural world, even in the minds of some scientists. This situation needs to be explicitly addressed. Philosophy has helped us learn to discuss consciousness. The previous chapter attests to that. But the party is over.

In a career as an engineer, never once has a philosophical category played a part in a design. I was never formally trained in philosophy. But I have read a lot. Engineers and scientists probably need more philosophy, with its discipline in logic, specificity in language and innate grip on argumentation. What scientists and engineers do not need is philosophy as explanation of the natural world. Science/engineering is a particular form of practical argumentation that is unlike philosophy. As a form of argumentation, the scientific experimentation process has its own problems. But these are unlike the problem of philosophical positions, such as XYZism, as a form of knowledge. XYZisms are merely opinions about the meanings of words. You can argue the meanings, and the outcome may be fascinating and engaging, but in the end they are just a bunch of abstract symbols spewed out in response to a prior stream of abstract symbols, which meet more in reply, and so forth. Meanwhile the natural world marches on, unaddressed.

> Scientists have not generally needed or wanted to be philosophers. Indeed, normal science usually holds creative philosophy at arm's' length, and probably for good reasons.

> [Kuhn and Hacking, 2012, Page 88].

As fascinating as they can be (if you immerse yourself in them), no philosophical position ever informed a review of my engineering or scientific choices. Evaluation of system performance involved no reference to philosophy. Nor do I ever expect it to. In science and engineering, designs are started, evaluated, built, tested and variously worked or didn't, and are abandoned or redesigned as needed. And all of this proceeded without any philosophical aspects. Indeed the great bulk of participant engineers and scientists wouldn't even know what a philosophical category is (thereby adding to the malaise). As merely opinions, and therefore as ephemeral as the shifting sands of the meanings of words, XYZisms are not something you can build. They are not a design. They are not a description of anything real. They predict nothing. They explain nothing. They solve no problem. Other than possibly as a way of stirring up a scientist's neurons in novel ways that may then end up practical for science, they are devoid of practical guidance.

Furthermore, I have no obligation to calibrate my investigative processes in terms of philosophical XYZisms. I certainly don't have to justify my choices to philosophers. Nor should any funding be contingent on the espousing of a philosophical XYZism. Empirical work speaks for me. Once the consciousness problem is solved, there will be a *post hoc* investigation of what philosophical XYZism categories the real solution seems to present. There can be a great analysis of who was right and wrong. If there are 50 -isms that all seem to have some kind of story to tell, but none are actually 'on the money', then some new kind of philosophical -ism will probably be created that aggregates the winning attributes of the old ones, and however interesting it may be, we scientists/engineers will remain as uninformed and guidance-impoverished by that as well.

In science and technology, the mere holder of an opinion devoid of potential empirical substantiation is, by definition, the loser of an argument that also loses the right to be involved any further in the argumentation process. This is the position of the climate-change denier in the current climate science situation. They have no argument because they have no evidence, and therefore should not be in the debate at all.

As opinions, nothing on offer by an XYZism has any basis in verified fact or is even testable because I could change the meaning of a word and refute the 'XYZism science' every time. Nor is an XYZism directed at making choices or otherwise resolving the 'argument' that is literally the completion of a technology or an experiment design. In any given project, if you lose the argument or have no argument, you simply remain silent until there is an argument (evidence) that may change a design or other policy. Philosophy's role in the realm of explanation of the natural world at the coal-face of scientific enquiry, and in particular, on Wheeler's beach, the shore of scientific ignorance, is therefore forfeit.

It is for this reason that I will not be considering any of the 'explanations' of consciousness provided by an XYZism, or its surroundings, as worthy of discussion or empowered to act in critique of the proposals in this book. In this book, everything that is proposed arose and is tested empirically. All argumentation starts and ends like that. Philosophy can help construct critical arguments and question reasoning, but it has no other basis to critique its content.

It has been rather strange to exit engineering and enter science to solve a particular problem, and find philosophical discourse masquerading as science. Imagine my frustration at the expectation that I must engage this discourse in some way prior to my acceptability as a contributor in certain areas of the scientific literature. My location in Figure 4.1 position (a) in the science of consciousness puts me in this very position. That is 'what it is like' to experience that very Kuhnian circumstance.

4.2 The taboo

There is good reason for the strange confrontation I claim exists between philosophy and science at this transitional juncture. It is obvious that I am converging on a direct link between a claimed log-jam in the science of consciousness and the lack of attention to a science of scientific behaviour. This is, I hope becoming gradually clear as the circumstances are spelled out bit by bit. That there is such a state of affairs is certain. A valid question is how the circumstance arose and is maintained.

The scientific explanation of consciousness has been fraught with centuries of controversy and exile in the realm of the science pariah. During this era of pariah-hood, if a scientist explicitly addressed consciousness, the hapless soul would be directed to the philosophy department, and that command was overtly pejorative; almost a definition of unscientific, with career-limiting (or certainly fund-limiting) implications. Something to be spurned with disdain. I have experienced this attitude myself, more than once, in the last 10 years. Usually it is exhibited by the older generation of scientists. Thankfully, as these people leave the field the attitude goes with them.

This says what happened. Why did it happen? The brute fact is that science was able to wrest away, from philosophy, all responsibility for the construction and validation of laws of nature by using the method shown in Figure 4.2(a). It works. Subject area by subject area has succumbed to the method. And the one thing it critically relies upon is empirical observation, which involves P-consciousness, something that was labelled very early on as an unexplainable thing, not because it was explicitly argued as fundamentally unexplainable, but because science objectified the observer out of the picture, thereby making it unscientific in a cultural/procedural sense. Scientific method developed objectivity, which is literally designed, for good reasons already discussed, to render a target of study independent of the peculiarities of person or circumstance lest it make a law of nature dependent on a particular individual observer. Observer-independent generalisations are the 'Law of Nature' order of the day, and the more universal the generalisation, the better. It rests entirely on ensuring that the subjective life of the individual scientist is objectified out of the process. Obviously you can't do a 'science' of a scientific observer if science is *defined* to objectify the observer out of the picture.

Is it any wonder that scientists of earlier times that exiled the study of the scientific observation process (P-consciousness) were rewarded by success? Holding a vehement objection to the study of subjective experience worked, especially in an era that had nowhere near the modern neuroscience, computing and complexity theory we now know is necessary to deal scientifically with consciousness. Curiously, the old-school scientists were more aware than us (here in the 21st century) that what they did was 'organise appearances' e.g. Ernst Mach and George Henry Lewes [Lewes, 1879; Mach, 1897]. Yet at the same time, so strong was the general paradigmatic objection to anyone studying consciousness, that 200 years of evidence in the literature demonstrates that only as scientists approach the sunset of their career, with nothing to lose, do they get out 'the sensations' and 'the given' and plaster their various late-career books with material. They did it because they could do it safely.

Here is an example from Ernst Mach, who realised that the sensations are all we have to access the world, and that our machinations as scientists are all about 'organising the appearances' that we get as a result. But then he also ran into a mental blockage that stopped him realising that we scientists are also evidence of the system of the production of the sensations (see Figure 4.3).

> Ordinarily pleasure and pain are regarded as different from sensations. Yet not only tactile sensations, but all other kinds of sensations, may pass gradually into pleasure and pain. Pleasure and pain also may be justly termed sensations. Only they are not so well analysed and so familiar as the common sensations. In fact, sensations of pleasure and pain, however faint they may be, really make up the contents of all so-called emotions. Thus, perceptions, ideas, volition, and emotion, in short the whole inner and outer world, are composed of a small number of homogeneous elements connected in relations of varying evanescence or permanence. Usually, these elements are called sensations. But as vestiges of a onesided theory inhere in that term, we prefer to speak simply of elements, as we have already done. The aim of all research is to ascertain the mode of connexion of these elements.
>
> Ernst Mach [Mach, 1897]

It's now possible to see how the Figure 4.2(a) method works brilliantly everywhere insofar as it creates descriptions (organises appearances or Mach's 'ascertaining the mode of connexion of elements') that are predictive of how the natural world will appear when we look. It is rather obvious now that the one place that the Figure 4.2(a) method can, well in advance, be predicted to fail, is a prediction of what

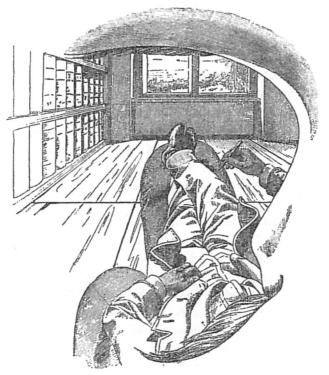

Fig. 1.

Figure 4.3.   The first person perspective from the point of view of Ernst Mach's left eye. Ernst Mach said "The assertion, then, is correct that the world consists only of our sensations" [Mach, 1897, Page 10]. This sounds bizarre. The world does not 'consist of' sensations! The world is made apparent to us through their provision, by brains, to us. The words Mach uses betray a mental blockage: that the sensations are given or a start point beyond which there is nothing. This blockage seemed to be endemic in that era. (Image courtesy of Open Court Publishing Company.)

the natural world looks like in the act of doing a scientific observation (it looks like a brain). The very existence of a scientific observer is outside the realm of prediction of the Figure 4.2(a) procedure. The observer is implicitly presupposed in Figure 4.2(a) and has been since day one of its usage.

The origins and continuation of the taboo is well recognised and documented [Wallace, 2000]. The remnants of the taboo have tentacles lodged in recent history. The further back in time one goes in, for example, the 50 years since Kuhn's '*Structure*', explicit scientific attention to consciousness apparently wanes and its funding was terminologically camouflaged, lest funders get wind of it. The birth of cognitive science, which happened around the time of Kuhn's book, was an early refuge for hidden attention to consciousness, and even cognitive science was held as non-science for a long time. But late last century that changed. As described in the previous chapter, since approximately 1990, *explicit* attention to consciousness, and a viable publicly funded empirical paradigm (the 'neural correlates' of it) has found its feet at last. It has already been shown to be logically compromised as 'explanation', and there will be much more on this in future chapters. But at least there is a science, and valuable progress has been made as a result.

The recent history shows us as having secured an empirical method using modern technology and then climbed back aboard the consciousness boat. That done, we scientists find ourselves on a tour boat to nowhere, with philosophers at the helm, created by centuries of our own paradigmatically implicitly self-enforced neglect. In the establishment of a direct link between the science of consciousness (the scientific observer), and the absence of study our own Figure 4.2(a) scientific behaviour, the taboo of consciousness is held responsible for an unmanaged abdication of responsibility for explicitly governing ourselves. As a result we have got the science we asked for, Figure 4.2(a), which, it will be shown more formally later, is now actively directed at the one thing that it was never designed to do – the scientific study of consciousness.

4.3 Summary

In this chapter the last of the relevant history was fleshed out. It was demonstrated how scientific behaviour has been, for several hundred years, a stable behaviour that has traditionally eschewed attention to consciousness. This aversion to consciousness took the form of a taboo that left the matter to philosophers. This era of stability of science practice fits the Kuhnian category of 'normal science' and has been extremely successful. As a result we scientists have been able to operate without attending to the science of consciousness, or even the possibility of one.

But now we have a science of consciousness and are currently in the unusual state of having to do what was last done well out of living memory: take, from philosophy, control of a natural world phenomenon previously outside the purview of scientists. At the same time as giving birth to a science of consciousness, we are applying, to the science of consciousness, an undocumented scientific behaviour paradigm that has been so stable for so long that we have no awareness that the same birthing process must have happened to science itself somewhere in the past. In the previous chapter we revealed that the nascent science of consciousness is in an anomalous state that challenges the validity of the undocumented paradigm in a way that no living scientist can possibly be prepared for because it is not written down and is trained by imitation. This is the next stage in the compounding account that is building our case for a change to scientific behaviour itself.

Chapter 5

'Normal' Science

The whole of science is nothing more than a refinement of everyday thinking. It is for this reason that the critical thinking of the physicist cannot possibly be restricted to the examination of concepts of his own specific field. He cannot proceed without considering critically a much more difficult problem, the problem of analyzing everyday thinking.

On the stage of our subconscious mind appear in colorful succession sense experiences, memory pictures of them, representations and feelings. In contrast to psychology, physics treats directly only of sense experiences and of the "understanding" of their connection. But even the concept of the "real external world" of everyday thinking rests exclusively on sense impressions.

Albert Einstein [Einstein, 1950, Page 59]

In Kuhn's *Structure* Chapters III (The Nature of Normal Science), IV (Normal Science as Puzzle Solving) and V (The Priority of the Paradigms), we see a lucid account of the paradigmatic norms of science *output*, and also witness the lack of scientific attention to regularity in the scientific behaviour that produces it. Kuhn is not alone in an inability to explicitly articulate scientific behaviour. Nor is he alone in his lack of awareness of the need for it. The literature up to that time shows (in my searching so far) that this lack of awareness is universal across all of science. Since then, there has been a gradual biological 'turn' in philosophy that one might think could have led to the biologising of an understanding of scientific behaviour. Yet it has not. Here is a recent example:

Science is not done by logically omniscient lone knowers but by biological systems with certain kinds of capacities and limitations. At the most fine grained level, scientific change involves modifications of the cognitive states of limited biological systems.

Philip Kitcher [Kitcher, 1993]

In a future chapter a more formal treatment of human cognition as a complex dynamical system will be used to set out a contemporary way of understanding scientific behaviour. Such a view is of practical value when considering how to make an artificial scientist. It is hoped that scientists might see themselves in such a discussion and that the discourse might migrate out of philosophy as a result.

5.1 Normal science as paradigmatic science

Without delving deeper into what a paradigm might be, we start by noting that Kuhn's depiction of paradigmatic scientific output from the historical literature reveals eras of overarching investigative regularity he characterises as 'mopping up'. This is not meant in a pejorative sense. Paradigms gain their status by being successful at solving commonly recognised acute problems, and the promise of success discoverable within an identifiable raft of under-addressed problems. A healthy paradigm is rich in such mop-up work, and this 'normal science' engages most scientists throughout their careers.

Normal science tends to find genuinely novel natural phenomena invisible. It tends not to invent new theories and its practitioners are intolerant of others that do. In the modern era, with the Internet, those on the novelty fringe outside the paradigm find each other and aggregate [Wertheim, 2011], yet remain outsiders. This is just a more modern symptom of the same 'normal' or 'received view' science paradigm. Normal science still can be said to operate with the kind of drastically reduced or restricted vision that Kuhn then argues is essential to the developmental depth and precision of intra-paradigm science.

The historical literature's presentation of normal science, Kuhn suggests, is exhaustively represented by three kinds of fact gathering that are present in both theoretical and empirical form:

(i)   Acquisition of facts shown to be particularly revealing of the nature of the things of concern to the paradigm. The more precision and the greater the variety the better. The most obvious recent example of this kind of activity is the human genome project, which successfully enumerated the entire nucleic acid sequence of one human individual.

(ii)   The acquisition of facts for comparison with paradigm theory. A recent and classic example is the 2012 acquisition of 'facts' by the Large Hadron Collider (the biggest single machine or instrument built so far), which was all about validation of the predictions of the standard model of particle physics, in particular the existence of the Higgs boson.

(iii) Acquisition of facts undertaken to articulate the paradigm theory, resolving its remaining ambiguities and addressing unsolved problems. This might include the determination of physical constants. It might result in the articulation of new, context-specific quantitative physical laws (theory-driven, not curiosity driven fact acquisition). It might include experimental application of paradigm theory to novel areas of interest, even in otherwise separate disciplines. Paradigm articulation is therefore the most obviously equally theoretical and experimental.

In the 21st century we have these three activities operating on an industrial scale, and have automated much of the routine, but the same three characteristics of science output remain in operation. None have changed fundamentally. No new kinds of output have been created. With that stability, one is entitled to ask what underlying natural invariant has applied all along. The answer defines the behaviour of a human when doing science that produces outputs of kinds (i)...(iii), and has not been delivered by any of the 'fact-finding' behaviours conducted so far. Nor is its acquisition the explicit mandate of any of the multitude of scientific instances of activities (i)...(iii), from low level physics through neuro/cognitive science and on into psychology, the social sciences and ecology. This is a search I have performed personally, perhaps not with the depth of Kuhn, but as best I can. Furthermore, we scientists remain, as has already been noted, unaware that this absence is a fact of our mainstream lives.

5.2 Puzzle solving as normal science

Even the project whose goal is paradigm articulation does not aim at unexpected novelty.

[Kuhn and Hacking, 2012, Page 36]

This is the way Kuhn questions the motivation of scientists, who can be seen in history as having failed as a scientist when they failed to produce a paradigm's expectations. If the result is prescribed in advance by the paradigm, why bother? Kuhn's answer to this is the motivation that arises in the art of puzzle solving as articulation of paradigm. The challenge behind scientist's motivation to involve themselves in paradigm-restrictive problems is a fascination with the solution of problems involving complex instrumentation, conceptual, mathematical and logistical puzzles. Consider the large hadron collider (LHC). If that is not an example of a gigantic puzzle solution with exactly these characteristics, directed at a known/expected result (the Higgs boson), then what is?

In puzzle solving, the scientist has the luxury of taking the paradigm for granted, and testing their ingenuity and skill against accepted problems. Problems that are not acceptable get marked as 'metaphysical/unscientific' or 'someone else's problem' or 'too troublesome'.

One of the reasons why normal science seems to progress so rapidly is that its practitioners concentrate on problems that only their lack of ingenuity should keep them from solving.

[Kuhn and Hacking, 2012, Page 37]

Not only is the science puzzle solver expected get a paradigm-compatible solution, the solver is expected to solve it in a particular way, so that the solution might self-communicate recognisably to the paradigm's community, who will then judge its adequacy based on the shared expectations of the community. We can now contrast this with the more romantic notion of the scientist that is embedded in a big picture of discovery, new territory, new order and the challenging of long held beliefs. The reality is that any individual scientist at the coal-face is actually not (obviously) doing any of these things, because the minutia of puzzle solving is what is transacted on a daily basis.

... These are explicit statements of scientific law and about scientific concepts and theories. While they continue to be honored, such statements help to set puzzles and limit acceptable solutions.

[Kuhn and Hacking, 2012, Page 40]

Kuhn characterises the working level of the individual this way, then adds:

Finally, at a still higher level, there is a another set of commitments without which no man is a scientist. The scientist must, for example, be concerned to understand the world and to extend the precision and scope with which it has been ordered. ... Undoubtedly there are still other rules like these, ones which have held scientists at all times.

[Kuhn and Hacking, 2012, Page 42]

This comes about as close as I have found to being the place, in the literature, where I commence the contribution by this book. Here we get to the nub of the issue of normal science. What rule might best categorise what scientists do? Kuhn puts us in the departure lounge for the upcoming journey. Yet it simultaneously implicitly denies there is such a journey.

In these and other respects a discussion of puzzles and of rules illuminates the nature of normal scientific practice. Yet, in another way, that illumination may be significantly misleading. Though there obviously are rules to which all the practitioners of a scientific speciality adhere at a given time, those rules may not by themselves specify all that the practice of those specialists has in common. Normal science is a highly determined activity, but it need not be entirely determined by rules.

[Kuhn and Hacking, 2012, Page 42]

Consider the phrase "*need not be determined by rules*". 'Need not' does not mean 'cannot'. Is Kuhn saying that it is impossible to capture the essence of scientific behaviour with rules or that we shouldn't? It is comforting to think, as a scientist, that somehow the essence of my humanity and my special brain faculties contribute an ineffable magic to the process of procuring 'scientific statements' ('rules') of the kind he discusses. The fact that we have not been encouraged to find the 'rule of scientific behaviour' does not mean there isn't one. And I for one do not need to hold my science-capable internal life sacred and unapproachable by science itself. In any other arena this 'rigidly defined area of doubt and uncertainty' would be cast as a religious tenet. Do we really have, operating at the heart of science, an implicit belief that our own behaviour is beyond scientific access? For that is the nature of the situation after a large survey of science and scientists. It suggests an implicit rule of the form *"Scientific behaviour (the extraction of natural regularities) shall be applied to everything in the natural world except scientific behaviour"*, operates permanently in the lives of scientists.

I came to science from engineering devoid of all such predispositions and I see no reasons for failing to extract rules that define scientific behaviour that is a rule about the natural world like every other rule we construct, especially when it is so easy to construct and so revealing when you do. Should such an explicitly stated rule or rules prove to be useless or irrelevant or otherwise objectionable, *then* we can say, as a community of scientists that *"we tried it and it was unproductive"* and have evidence to support it. Not before. We haven't tried yet. This book makes the first attempt, and let it be the first round in an explicit ongoing dialogue.

Now take a look again at the previous chapter, where we discover a centuries-old (albeit abating) taboo in science: consciousness (a science thereof). Historically, the idea of a science of consciousness was clearly not regarded as mere puzzle-solving within an existing paradigm. If ever there was a topic regarded as paradigm-toxic, this is it. Until the last half of the 20th century, and most likely the last 20 years of it, the scientific study of consciousness has been uniformly set aside under the Kuhnian 'metaphysical/unscientific' category. The reality of the existence of a science of consciousness means that where once it was unscientific, now it is scientific. Where once it was mere 'metaphysics', discarded to the realm of armchair philosophers, now there must be a sense in which it is not. Otherwise how can that science proceed? The answer is that it has been 'normalised'. To do the science of consciousness is to find the 'ABC-correlates of consciousness'.

This is an extraordinarily telling position for science to be in, and much will be made of it later on. It suffices to say for now that science has allowed itself to dance around the fire of consciousness, but to remain in a position of making the same kind of statements – presupposing the observer – that have always been made. We have thereby enabled the old paradigm to do something in an area that was previously forbidden ground. We scientists get to be scientifically 'normal' but are (a) revealed as operating in a manner devoid of explanation of the studied phenomenon and (b) logically compromised by using scientific observation as some kind of explanation of scientific observation. This is the grand tautology of the 'ABC-correlates of consciousness' empirical paradigm. It is descriptive, predictive in some expected sense, but is devoid of actual explanation. Think about it. You plumb the depths of physics in the brain, you find the perfect correlate of a REDness report, you can control it, reproduce it ... and yet you have no clue why it is that the owner (the subject) should have a first-person perspective at all, let alone REDness, as the result of bits of the brain dancing about in a way that you know, scientifically, everything about.

Furthermore, we got to this position by doing exactly what Kuhn says we have always done: accounting for the behaviour of the natural world by articulation of paradigm theory, which in this case means retaining our implicitly enforced confinement to treating observation as description, and of not explaining the observer. We have literally allowed ourselves to directly look at how it is we scientifically observe (with consciousness) by renaming it something else (the ABC-correlates of consciousness) so that we avoid having to admit that we are actually studying the scientific observer, something previously taboo. So we haven't kept science 'normal' by adhering to any explicitly stated collection of 'laws of nature' defining the purview of any explicit community in science. For the first time in history we have 'normalised' a science by ensuring it fits with scientific strictures that are not actually written down anywhere. We can say we are articulating paradigm theory again, but the theory is one that is not written down. It is a scientific theory about what scientific behaviour entails, that is implicit, tacit and passed on by imitation. This seems significant to me, and in need of further elaboration.

5.3 Paradigms and scientific behaviour as tacit knowledge

Tacit knowledge is difficult to transfer to another person by means of writing it down or verbalising it [Polanyi, 1967]. Kuhn recognised tacit knowledge as capturing the essence of paradigmatic scientific behaviour. In contrast, formal knowledge can be transferred explicitly. Scientific statements about the natural world ('Laws of Nature') are explicit knowledge. Deployed to engineers, they can be used to create technology without the involvement of the scientists that created the statements. An ability to play tennis (the physical prowess) cannot be conveyed explicitly. It is tacit knowledge. To construct this knowledge you must play tennis. The rules of tennis are explicit knowledge. The rules of tennis act as a set of constraints through which one can formally detect whether a game of tennis is being played. Language itself operates the same way. It is impossible to verbalise exactly how it is that verbalisation is possible. We just do it, and we cannot help learning to do it. It is built in to our physiology to be able to soak up such tacit knowledge. Modern neuroscience would classify tacit knowledge as being stored 'procedural memory', in contrast to 'declarative' memory of specific facts or episodes. The former is 'reported by doing'; the latter is reported through communication (e.g. verbal).

By direct analogy, scientists are trained by 'doing' (which includes learning laws of nature, mental exercises/problem solving as well as laboratory work; i.e. theory and practice) and just like the prowess in playing tennis, the scientist ends up with a collection of tacit knowledge that enables the production of new laws of nature and problem solving within a domain of specialisation. In a very real way, scientific behaviour may be described as 'sciencing' – a verb describing the act of it in the same way that playing tennis describes the activity.

So what exactly are these scientific paradigms? Are they what we (as scientists) might intuit them to be? In *'Structure'*, Kuhn, in his general psychologising of science, linked paradigms to tacit knowledge of the same kind that results in the act of explicit classification of concepts such 'chair', 'leaf', 'game' [Kuhn and Hacking, 2012, Page 45] or better, 'force', 'mass', 'space' and 'time' [Kuhn and Hacking, 2012, Page 47]. When we 'know' these things it is unequivocal and without argument. Behind this reality is mental dynamics that tacitly classifies something has having a resemblance (Kuhn called this *"a network of overlapping and crisscross family resemblances"*) to its categorical conspecifics, then delivers a result that we verbalise or act in accord. A paradigm is just such a set of conspecifics, but applied to scientific behaviour across a disciplinary subgroup (community); its outputs, its problems, its techniques and its instrumentation. We know it as a paradigm when we are operating in it in the same way we know a leaf when we see a leaf. An account of how we know is part of as yet unspecified, unknown neuroscience.

Kuhn's book is all about the dynamics of paradigms which, by their very nature, claim their community members through tacit knowledge expressed in the scientific traditions of an era and that are passed on in the process of professional initiation (the PhD process). Compounding the problem of the lack of explicit identification of paradigms is the fact that, even in 'normal' science, two individual practitioners from different communities address the same natural world through the lens of the respective community and thereby can see a quite different thing, yet both operate legitimately as scientists because their paradigm is different. Both add to the body of accepted scientific knowledge. Individually, however,

> Scientists work from models acquired through education and through subsequent exposure to the literature often without knowing or needing to know what characteristics have given these models the status of community paradigms.
>
> [Kuhn and Hacking, 2012, Page 46]

Remember, it is an historian that is retrofitting the characteristic 'paradigm'. It is not something the scientist is allocated or is even aware of in their encounters with legitimate scientific problems.

> That scientists do not usually ask or debate what makes a particular problem or solution legitimate tempts us to suppose that, at least intuitively, they know the answer. But it may only indicate that neither the question nor the answer is felt relevant to their research.
>
> [Kuhn and Hacking, 2012, Page 46]

Kuhn then describes the paradoxical state of scientists able to talk easily and well about current research, yet behave as unqualified laypersons when characterising how their research forms part of an established field as a legitimate problem, solution or method. This is like asking a tennis player to articulate the exact role of musculature used to carry out a serve. The musculature is presupposed in exactly the same way a community's established methods, problems and solution base is presupposed. While operating normally no rationale is necessary and through their relative rarity, a scientist need not be trained to handle anomalous/crisis science. Paradigm-change expertise thereby tacitly absents itself from the life of a scientist. Nevertheless we can explicitly portray some characteristics of the paradigm. For example, in their birth, Kuhn writes

> To be accepted as a paradigm, a theory must seem better than its competitors, but need not, and in fact never does, explain all the facts with which it can be confronted.

> [Kuhn and Hacking, 2012, Page 18]

You might have the impression that a paradigm is merely a theory or a collection of theories. Far from it.

> Paradigms may be prior to, more binding, and more complete than any set of rules for research that could be unequivocally abstracted from them.

> [Kuhn and Hacking, 2012, Page 46]

The dynamics of paradigms include their demise, death and rebirth. Their demise occurs when the established traditions are challenged to the point of becoming insecure. Only then do scientists come to question the rules of their paradigm. This becomes necessary because somehow the long-held methods, problems and their solutions are under threat. These are the extraordinary problems that emerge only on special occasions prepared by the advance of 'normal' research that finds it can only be conducted in an abnormal way. In trying to understand what has gone wrong, scientists attempt to articulate their paradigm (rather than the paradigm's theory), in which case Kuhn advises that this has the effect, during paradigm shift, of dividing a community into schools, rather than addressing the offending natural phenomena that created the problem.

5.4   The 'Law of Scientific Behaviour' – the first steps

We reach this point, I hope, having more formally established a working knowledge of 'normal' paradigmatic science and the rudiments of paradigm dynamics. We have Kuhn's history describing the way scientists behave in terms of the alterations to scientific *output* – the products of science, not the physical behaviour that produced it (the natural world of the behaving scientist). I have found almost nothing has changed since Kuhn's book in 1962. We now have more developed theories of learning, including tacit knowledge. In addition we now recognise (as per the Kitcher quote above), that the basis for science is brain dynamics understood from a neuroscience perspective. Regardless of these modern developments, it remains basic that the paradigmatic behaviour of entire science communities, all operating in the absence of an explicit definition of their hosting paradigm, has not changed since Kuhn depicted it 50 years ago.

It also remains the case that scientific behaviour is acquired without passing on any formal definition of what it is. So it is time that we examined what such thing might look like. We seek a universal 'law of scientific behaviour' or a 'law of nature about the establishment of laws of nature'. First, note that paradigms come and go, but what remains invariant is the scientist. In particular the cognitive faculties that enable the 'three kinds of fact gathering' that Kuhn says exhausts the list of scientific activities. All these activities ever do is result in creation, modification or discarding of statements by scientists. Although at the moment they are produced by tacit knowledge, laws of nature are obviously not tacit knowledge. By definition they are explicit and they are of a kind produced by a specialised community of humans behaving in a unique way. Their production and change is a natural regularity that outlives any and all paradigms and began when humans first decided to explicitly capture regularity in the natural world.

Grunts and hand signals and marks on a stick would do the job of making the knowledge explicit. Then, as the behaviour stabilised, scientific behaviour itself became another regularity in the natural world. It's impossible not to. That regularity can be scientifically captured. A statement about the regularised production of statements of natural regularity is a member of the set of all statements of natural regularity. As soon as human scientific behaviour acquired regularity, that behaviour itself became a suitable scientific target. The fact that we never actually did that is just a matter of history.

Scientists construct statements and these statements capture regularity in the natural world. There is no need to claim anything esoteric or difficult about the nature of the statements or their status as truth. The philosophical treatments of epistemology are irrelevant. Once created, the statements acquire their power through the community-wide (tacit) knowledge of the language that mediates the communication and subsequent use of the statements. Of course, only those folk able to internalise, understand and act on the statement give the statement viability as a predictor of the natural world. This technicality acknowledged, we can then realise it is only when scientific statements inhabit the head of individuals that they acquire their intended power: to be predictive of the natural world. This is the basic way Einstein's 'everyday thinking' becomes scientific behaviour.

The second invariant across and through all paradigms is that, regardless of the statements produced, it is recognisably science that is being done. These are scientists, not lumberjacks or ballet dancers. Their behaviour is very specific. At the end of the day they are all participating in the production of statements that are regarded as valid by the community's paradigm. As each day passes, one can argue that an expanding body of statements (Wheeler's ever-growing 'island of knowledge') has had some attention, and that the activity surrounding it has unique characteristics that separate it from other human activities.

All human activity generates knowledge of some kind. In the case of scientific knowledge production it is largely the target (nature), the explicitness of it and the portability (across humans) that separates scientific behaviour from other human behaviours. Acceptance of a scientific statement by the community, for our purposes, does not mean these statements are right or 'truth' or even sensible. Indeed, in antiquity, statements are barely recognisable as science, yet were held as forcefully, arose from paradigmatic methods, were argued as vigorously and those that generated them would be labelled scientists of the time. Statements are, for the purposes here, merely things accepted by a community that stands in judgment of them on their predictive merit. Other statements may be rejected. Changes in paradigm occur when some long-held subset of such statements is invalidated in some important context.

It is interesting to note that humans can make technological advances without any science of the kind described above (explicit statements). Animals such as chimps and birds have technology: *tools*. The animals have technology in the same way that proto-humans had technology (such as fire). The knowledge itself and its means of acquisition and use, are tacit and tied up in basic neurological processes that result in technology creation and usage. There need not be any trading in explicit abstractions of the natural regularity behind the tools. Nevertheless, the technology can come into existence and the knowledge of it can be passed on, through imitation, as usual. All of this can happen without the existence of any scientific behaviour of the kind important here and that is enacted by modern humans.

Before proceeding we need a little more technical specificity in the idea of a 'statement'. Let's identify any statement made by a human that might be put forth as the output of an act of scientific behaviour. Whether it is predictive or not, denote the statement using the following format:

$$t_n = \textit{Some statement about the natural world.} \qquad (5.1)$$

As soon as we do this, we recognise the tacit knowledge state of the utterer of such a statement. The utterer has oral/written language skills. The statement itself can be total rubbish such as $t_{1000} = $ "*The volcano is angry and we must make offerings to appease it*". Despite its apparent strangeness in a modern context, t_{1000} is a *belief* - a configuration of brain material that resulted in the communication of a belief on the part of the author of t_{1000}. These statements are not claimed to be statements of 'fact' of some kind. They are presented as an explicit reflection of the brain dynamics of the utterer – a belief embodied in the believer's brain subsequently made explicit in writing. How this works is to be detailed in a future chapter. Nothing more is claimed about the status of any t_n of the equation (5.1) kind. As science, the predictive effectiveness of t_{1000} rests on the natural correlation between offerings and volcano eruption. That state of affairs is moot to the discussion.

Here's a 'law of nature' that was made popular in the 17th century and in various incarnations can be applied to this day: $t_{1001} = $ "*f=ma*". Newton's second law could use a little embellishment, but I think you get the idea. Once again, all that is claimed here is that it is a belief, and when inhabiting the brain of a human with appropriate tacit knowledge (mathematical/language/science training), the believer gets to predict the natural world in the context addressed by t_{1001} (which you will note is not yet explicitly mentioned in t_{1001}). Note that the language of science in those days was Latin, which tended to limit the extent to which t_{1001} could take up residence in the brains of the general population.

Now we have the generic form of any scientific statement, we return to our statement about the 'natural world of scientific behaviour'. With our new nomenclature we now know that a law of nature depicting the act of acquisition of laws of nature, even though it is confusingly self-referential, also has the form of t_n like any other scientific statement. Because it is rather special in the scheme of scientific laws, let's call it t_A. It sounds ridiculously trite, but the easiest universal invariant is the statement t_A = "*scientists make statements about the natural world*". It is of little use, except as a stepping stone to more useful versions. It's a bit like saying "*tennis players play tennis*". If the natural world was devoid of the game of tennis this statement would be meaningless. Note, however, that it may actually be that someone made the statement "*tennis players play tennis*", as gobbledygook, prior to the existence of tennis. It acquires meaning after tennis exists. This is more revealing of the reality of science than it initially appears.

Consider the gobbledygook version of a t_n such as "*it is a property of the natural world that X applies when some event happens*". This may be completely hypothetical and false at the time of the statement, making it indistinguishable from the above example of tennis related gobbledygook. Yet later, when the state of affairs at which it is directed is verified, it becomes a scientific statement. If not, then it is discarded. This is the practical nature and role of the hypothesis and it is revealed as having undefined truth status (gobbledygook) at the time of manufacture. Scientists make these kinds of statements routinely. Indeed, we must make them so that we can test them against the natural world. Gobbledygook becomes a law of nature through scientific behaviour. As a community-sanctioned belief about the natural world, a hypothesis is, initially, officially false. It is not, at that time, a member of the set of all laws of nature. Notice what this process requires of the mental faculties of the maker of statements: *it requires us to be innately capable of uttering things that are of doubtful status as truth (or even total rubbish)*. This leads us to the somewhat paradoxical realisation that in order that scientists access some kind of 'truth' about the natural world in the absence of *a priori* knowledge, *we have to be able to be wrong*. Try it: $t_{1002} = (1+1=4783547)$. This is wrong as a statement of base ten integer arithmetic, yet quite naturally utterable by any human.

Once again, remember there is no need to burden the status of these statements with status as some kind of fact or truth or building block. All you have to entertain is that they are statements that make you predictive to some useful extent. Nothing beyond that needs to be claimed a property of a t_n. When a t_n is empirically tested, all you are proving is that it made us predictive. No other or extra status is proved as a result of the testing.

Now lets us consider the set of all such statements that are accepted as scientific at some point in time. Call it the set T. It covers the whole of science for all time.

$$T = \{t_A, t_1, t_2, \ldots, t_n, \ldots t_{N-1}, t_N\} \tag{5.2}$$

There is a raft of observations we can make about set T. First, set T is dynamic. Its contents fluctuate as scientific communities add, remove and modify its contents. Notice that set T is of size $(N+1)$. This is because it contains t_A, which is of identical form to any other law of nature, t_n. One can also note that as science specialisations proliferate, that the size of T, $(N+1)$, increases overall. It was empty until the first explicit scientific act took place sometime in our early linguistic history.

Next notice that the Kuhnian account of history tells us that one of the main characteristics of the contents of set T is that on occasions, there are mass extinctions and periods of great instability. Kuhn also tells us that there are long eras where 'normal' science cautiously augments set T along thematic lines, carefully confronting them with nature in the manner of the three 'fact-gathering' scenarios listed above.

But there must also be another set to hold interim 'laws of nature' that have not yet made the grade. These are the hypotheses that have an undefined status as a viable belief, and may as well be regarded wrong, or at best not predictive. To hold these let's define another set H:

$$H = \{h_A, h_1, h_2, \ldots, h_n, \ldots h_{N-1}, h_N\} \tag{5.3}$$

These set members have the same form as t_n, but are not valid 'laws of nature'. Set H might be said only to contain hypotheses, but because hypotheses are members of essentially failed laws of nature (they will never get into set T unless tested), in principle set H contains all statements that are hypothesis plus failed laws of nature. For example, set H contains the much discussed 'phlogiston' theory of combustion, and perhaps 'phrenology', 'faith healing' and so forth. The list is large but finite.

We can now see the dynamical process of science as set membership fluctuations. By following t_A, scientitific creativity populates set H, with statements h_m of type t_n. These are then tested, and become accepted statements in set T or they remain rejected statements in set H. Set T members can fail when counterevidence prevails, in which case they go to set H. All of this is entirely consistent with the Kuhnian depiction of the historical behaviour of scientists, and the punctuated equilibrium of paradigm and paradigm shift.

The interesting thing about set T is that until this book, set member t_A was invisible. Indeed, in the form I have constructed so far, its membership of set T appears to contribute little value. But consider how it got there. We have centuries of modern science, in which the set population behaviour was a natural regularity not in set T. Yet science arose all by itself in its absence, and when people like Thomas Kuhn looked at science, they never picked up the fact that it was missing.

Admittedly that absence can hardly be claimed important if t_A has the rather trite form given above. But I will later show the more accurate form of t_A, which is far from trivial and is very revealing. What is important at this stage is that science can arise, by virtue of the cognition of humans, and then begin populating set T without the need for an explicit t_A, through its presence all along as 'tacit knowledge'. The set-theoretic approach is also compatible with all the human eras of science, including authoritarian, superstition and religious eras. This is the power of treating set T and set H merely as a repository of explicitly expressed beliefs, and not enforcing any notion of truth or projecting any relationship with nature beyond that of being predictive.

We can now embark on further adventures with paradigms and the better elucidation of t_A. Before this, consider a more pictorial representation of the dynamical process of science as shown in Figure 5.1, where we see a scientist operating according to t_A evaluating a law of nature that may have come from the discard pile (via route d from set H, a hypothesis) or from set T (via route b) because there is another test being performed on a set T member. As a result of the activity, the statement goes to set T (route a) or H (route c) as appropriate.

Figure 5.1.   The basic dynamics of laws of nature.

We can now finally see what I have previously claimed is passed on by imitation from mentor to novice, to become 'tacit knowledge' in the novice: t_A and the Figure 5.1 context of its use. In a different world, the training would involve, amongst all the other aspects of scientific initiation, explicit exposure to t_A. At this stage it is hard to see what is being missed or under-examined as a result of an implicit, imitated t_A, and the implicitness of t_A is hardly going to be a source of error in the day to day life of a scientist. However, later on, with a real-life t_A fully articulated, it will be shown that a great deal is missed or not adequately understood by not explicitly capturing and managing it.

Next consider in these set-theoretic terms what a Kuhnian paradigm actually is. If one were to record the dynamics of set T over time, one could literally plot the stability of individual statements along with the appearance and disappearance of statements. Against a background of ever increasing set T membership, there would be brackets of times of incremental steady change. These would be the paradigms of normal science. The period of crisis, if recorded at fine enough resolution, would be revealed in set T instability. Discontinuities in this stability metric would indicate sudden alterations and the commencement of a new paradigm. This is where the previously mentioned 'power law seismology' would be observed in set T dynamics. Had science been conducted this way, this is what the history of science would look like.

Neither an accurate form of t_A, nor the proper generic form of laws t_n has been delivered yet. This was deliberate because many important attributes depend merely on their existence, not on their specific form. Their accurate form will come later. The fact that it is possible to make such statements is the key outcome. Also, scientists do not actually produce them in any standardised form. In reality we have a massive literature in which our scientific knowledge is dispersed in no particular form. Nevertheless, should we decide, we can access any small chunk of knowledge (like we did for Newton's second law) and recast it the form of a statement of kind t_n.

Before we get into the details of the real t_A and t_n we need to take a deeper look at the primacy of P-consciousness in scientific behaviour.

5.5 Summary

This chapter introduced the idea that current scientific behaviour is a very specific behaviour that can be scientifically described like any other aspect of the natural world. The nuts and bolts process of scientifically describing scientific behaviour was spelled out in preliminary form:

(iv) A set T of all scientific statements.

(v) A generic form t_n for any scientific statement in T.

(vi) A set H to hold untested (hypotheses) and failed t_n.

(vii) A special instance of t_n called t_A to describe scientific behaviour.

(viii) Together these will be referred to as the $t_n/t_A/T/H$ framework.

These were established without their full details, which will be covered in Chapter 8. The chapter also reinforced how science done as per the 'law of scientific behaviour', t_A, could arise and operate quite happily (as a describer only) without formally documenting our own behaviour. The concept of the 'scientific statement' was introduced. Scientists make statements, where the intent of the word statement is 'utterance' and the only claim about the statements that scientists can empirically justify is that they make you predictive (a describer). In this way we are gently introduced to the idea that science, as it is currently configured (and applied to the science of consciousness), is entirely neuroscience-centric. When 'laws of nature' change all it means is that human scientists 'change their mind' about which statements to apply to predict natural world appearances.

Kuhnian 'normal science' is witnessed in steady increases in the size of set T. Kuhnian crises and paradigm shifts are witnessed in the sudden large changes in set T. If there is more than one way to do science (one is constructed in Chapter 11) then there will be an alternate to t_A in T. It just happens to be missing at the moment.

Chapter 6

The Great Blockage

It is as elementary prototypes for these transformations of the scientist's world that the familiar demonstrations of a switch in visual gestalt prove so suggestive. What were ducks in the scientist's world before the revolution are rabbits afterwards. The man who first saw the exterior of the box from above later sees its interior from below.

....

Therefore, at times of revolution, when the normal scientific tradition changes, the scientist's perception of his environment must be re-educated – in some familiar situations he must learn to see a new gestalt. After he has done so the world of his research will seem, here and there, incommensurable with the one he had inhabited before. That is another reason why schools guided by different paradigms are always slightly at cross purposes.

[Kuhn and Hacking, 2012, Pages 111/112]

As part of the theme of current scientific behaviour as a purely neurological phenomenon, this chapter examines the nature of the mental shift needed for science to encounter the proposed changes to science itself, integrate and then normalise consciousness as a natural part of science. In broaching these issues it is hoped that some scientists may identify with their own mental inertia. The material may help to encounter the later propositions. Those more concerned with technical issues can skip to the summary.

If 'seeing' the new world of a traditional paradigm shift is difficult, then imagine what it will be like for us scientists to see our own objectivity and self-knowledge shift in a similar way. There is a reason why the proposed changes to science (which are not particularly hard) have not happened already.

6.1   Scientific behaviour and world-view gestalt

Chapter X of Kuhn's *Structure*, 'Revolutions as Changes of World View' is probably the singular place in which we can capture the essence of the difficulties to be expected when encountering my proposed change to science for the first time. Kuhn describes scientific worldview shifts as a kind of gestalt, like a Necker-cube or a duck-rabbit or a vase-face (Figure 6.1). Kuhn saw how post-revolution scientists literally saw the same natural phenomena (e.g. same retinal imprint) in a different way. Or perhaps in one individual the gestalt switch operates, and the seeds of scientific revolution are thereby sewn.

Figure 6.1.   Common gestalt illusions.

Fifty years of psychological research on perception makes this gestalt expectation even more plausible. It's now routinely understood and experimented on in a multitude of areas. Consider the 'rubber hand' haptic/pain illusion, where the subject literally integrates a fake hand into their own body image, and then projects, briefly, a pain response when the fake hand is hit with a hammer. Consider hemispheric rivalry illusions (visual version in Figure 6.1), where word selections and other perceptions depend, to some extent, on the dominance of the left or right hemisphere when the brain constructs a P-conscious sensory field. Consider the McGurk effect, where facial behaviour primes auditory perception, thereby literally causing the wrong word to be heard. We now understand that magic-trick illusions rely entirely on this kind of perceptual priming and the misdirection caused by what is now known as inattentional blindness and change blindness.

All these things and many more are now understood and routinely figure in the world of many scientists. The literature reveals the scientists that routinely encounter these phenomena are more at the neuro-cognitive end of the life sciences, and distanced from the basic physical sciences. This, it will be shown later, is one of the claimed factors that has inhibited its impact on science's own account of itself.

Kuhn uses the less developed 1962 knowledge of gestalt phenomena to great effect in understanding scientific change. He used the famous Aristotleian ⇔ Galillean revolution in the understanding of motion as an example. Both observed (their retinal imprints, say) or imagined some kind of swinging mass. Aristotle 'saw' a constrained falling rock. Galilleo 'saw' a pendulum. It seems perfectly natural for us now to impute a role for visual perceptual priming via language. The language system, now understood as tacit knowledge, highly integrated with and resonating associatively with the visual system, biased Aristotle's science statements towards one account, and Gallileo's to a different account.

This idea seems almost mundane now, except that when writing an account of the role of perception in scientific behaviour, I will show how *even Kuhn was subject to exactly the effect he was describing*. And it is precisely this effect that I see as behind the problems that I have had in bringing the material of this book to light. The fact is, that in scientists in the research/academic world right now, and even in young scientists, when P-consciousness is invoked in the role I invoke here, more than likely they stare blankly and have a response of the kind "*Huh?*" P-consciousness is a foreign concept in the mainstream scientist.

Now consider my personal position in this. I arrive in science after a career in practical engineering. I have a huge experience, but of a very different kind to the long-term captive of academia. I bring a different kind of tacit knowledge to bear on the problem of a scientific account of consciousness (as a problem with scientific behaviour, not a problem with an immutable, fixed scientific behaviour's output). I have been able to do this because I do not have the tacit knowledge of the kind that causes a systemic bias on one's mental picture (or lack thereof) of P-consciousness. As a result, exactly the kind of gestalt switch is at work. It is precisely this particular instance of gestalt switch that I claim biased Kuhn's words, and it is to that I now turn.

It's all in the words. When Kuhn refers to what we now call P-consciousness, he literally uses the phrase 'the given'. It sounds innocuous enough and rather quaint, even poetic. But it has another side. Kuhn was trained in physics, a highly mathematical science. In mathematics, 'the given' is an axiomatic point of departure for a mathematical proof. There are no priors to 'the given' axioms. Search the physics and mathematics literature. You will find endless use of sentences of the kind *"Given _____, it follows that _____ and therefore....."*. You can alter the axioms and explore the logical consequences, but that is all. Axioms are presupposed. This is one important part of the tacit knowledge priming the perceptions of Kuhn when 'the given' appears in his text and presumably elsewhere in the literature. There is no prior to 'the given'. Nevertheless, with great insight, Kuhn successfully isolates the problematic role of 'the given' in science. Consider the following quotation

> But is sensory experience fixed and neutral? Are theories simply man-made interpretations of given data? The epistemological view-point that has most often guided Western philosophy for three centuries dictates an immediate and unequivocal, Yes! In the absence of a developed alternative, I find it impossible to relinquish entirely that viewpoint. Yet it no longer functions effectively, and the attempts to make it do so through introduction of a neutral language of observation now seem to me hopeless.
>
> ...
>
> The operations and measurements that a scientist undertakes in the laboratory are not "the given" of experience but rather the "collected with difficulty." They are not what the scientist sees – at least not before his research is well advanced and his attention focussed. Rather, they are concrete indices to the content of more elementary perceptions, and as such are selected for the close scrutiny of normal research only because they promise opportunity for the fruitful elaboration of an accepted paradigm.
>
> ...
>
> As for a pure observation-language, perhaps one will yet be devised. But three centuries after Descartes our hope for such an eventuality still depends exclusively on upon a theory of perception and of the mind.
>
> [Kuhn and Hacking, 2012, Pages 125/126]

Notice the way that scientific output has been described in my previous chapter could not be any more in agreement with the apparently three centuries old view that 'theories are simply man-made interpretations of given data' (descriptions). I didn't call them theories. I just called them 'statements' because the word 'theory' is, in itself, too 'theory laden' (this I also claim to be a misdirection operating behind the scenes in Kuhn's text).

But this is a relatively minor matter. Of more concern here is the idea of a 'pure observational language'. The very existence of the idea of such a language is, I contest, symptomatic of the problem. The idea of it is a century old and goes by the name of phenomenology. It acts as if 'the given' are an impenetrable wall upon which is written the axiomatic 'contents of P-consciousness' that is all that we have, and that describing these contents in some standardised, neutral way can be a route to better science.

Because of my own tacit knowledge and perceptual priming in relation to this matter, it is a complete mystery, to me, how anyone could think that abstracted contents of consciousness recast in some sort of 'pure observational language' can be even possible or plausibly useful. Firstly, that ideal language is simply impossible in principle. It's like saying that there is a pure carbon that is not an isotope of it. Carbon exists in isotopic form. There is no 'official' or 'pure' carbon. The idea of a pure observational language of description is logically flawed in exactly the same way. Instead of functioning as a perceptually non-biased ideal, it would do the reverse. It would become some kind of proxy tacit knowledge and would force perceptual priming in line with the constructs of the language. How are we any better off? This is a deeply flawed idea. It is also deeply flawed in the terms of dynamical systems theory (Chapter 9) where we find that the same linguistic tokens (of this putative pure observational language) surf an infinity of different brain state-trajectories. A pure 'observational language' cannot guarantee identical trajectory traversal in the unique mental landscape of an individual user of it. The very idea is oxymoronic and inconsistent with modern neuroscience and dynamical systems theory.

But a deeper problem is why a 'pure observational language' seems plausible in the first place. I can't see how anyone would even begin to think it possible. That is how the gestalt in me, and the gestalt in phenomenology proponents speak from different worlds. Somehow the phenomenologist's mental world sees phenomenology as possible. I see it as a bizarre impossibility. Kuhn says I should expect this. Why should I consider my view better founded? Well, in my defense, I have neuroscience and dynamical systems theory that identifies the phenomenologist's presuppositions and that suggests that phenomenology is simply a faulty idea if applied to an account of the origins of consciousness or even in standardising scientific observation. In terms of the previous chapter, I deposit it into set *H,* which is what Kuhn did also.

I am more interested in why 'the given' can be seen as such an axiomatic impenetrable wall in the first place. It is evident (to me in my gestalt!) that Kuhn cannot 'see', in the very gestalt sense he is discussing, that 'the given' are scientific evidence of their origins, *in and of themselves,* regardless of their content. Above, I described the contents of P-consciousness as being 'writings on a wall'. I see a wall! P-consciousness evidences the 'wall' that presents them. In my world-view they are not 'the given' at all. There is a wall creating them, and that wall has origins.

Objectivity brilliantly extracts agreed contents of P-consciousness for scientists to use in elaboration of regularity-statements regarding the natural world outside the P-consciousness of any individual scientist. The natural world happily complies (within their agreed context/scope of applicability), ensuring that observations of nature will not be found to be inconsistent with the statements. But this very act makes the science critically dependent on the *existence* of P-consciousness in the first place. This critical dependency is meant in a literal, causal sense. Take the scientist's P-consciousness away and the entire enterprise grinds to a halt. It is as if 'the given' is a mental blindspot, and the fact that there can be no prior to the given is operating at the level of tacit knowledge. There is no awareness even of the possibility of a natural process by which 'the given' might arise, because it must be 'given'. Instead of accepting 'the given' as a natural phenomenon with a scientific explanation, Kuhn sees it only as something that has to be worked around, and then that somehow by proceeding as we always have, a theory of perception and of the mind is assumed within the eventual grasp of science as it has always been conducted.

This is the real taste of paradigm water and the pointy end of the presupposition stick. This is the blind that pervades science: it has 2000 years of tacitly presupposing a role for P-consciousness as originating scientific evidence, and 2000 years of setting aside P-consciousness via the use of objectivity and mental projections like 'the given'. And all of it constitutes a kind of uber-paradigm that has been there all along, through all the scientific revolutions, as tacit knowledge. A scientific account of the natural world has, all along, been defined to presuppose P-consciousness, and despite a very advanced appreciation of the role of human mental life in science, Kuhn is as much a part of that presupposition as anyone. That presupposition has the entire science enterprise 'organising appearances' while unaware that a scientific explanation of how 'appearances' arise at all challenges the very basis of the behaviour.

6.2 Manifestations of blockage #1 — science

In their book 'A Universe of Consciousness' [Edelman and Tononi, 2000], in the first and last chapters, authors Gerald Edelman and Giulio Tononi provide excellent examples of paradigmatic tacit knowledge, gestalt influences and the challenges surrounding the competition between a new and an old paradigm.

I have no great problem with the authors' basic technical propositions for brain operation and information integration, which I will not elaborate here. From my perspective their technical proposition could be right insofar as it characterises the nature of brain operation, yet cannot be an 'explanation' of consciousness and never will be. They think they are solving the problem of consciousness, when I can see that they are not. In my 'gestalt', the problem originates in a major disparity between our views of what constitutes science and scientific behaviour. Our different perspectives are immediately visible, even in the section headings. Let's begin with their chapter 1, 'The Special Problem of

Consciousness ...The Conscious Observer and Some Methodological Assumptions', where we find:

> Is a satisfactory scientific account of consciousness thus forever out of reach? Is there a way to untie the world knot? Or is there a way to break through both theoretically and experimentally to resolve the paradoxes of conscious awareness? The answer, we believe, lies in recognizing what scientific explanations in general can and cannot do. Scientific explanations can provide the conditions that are necessary and sufficient for a phenomenon to take place, can explain the phenomenon's properties, and can even explain why the phenomenon takes place only under those conditions. But no scientific description or explanation can substitute for the real thing. We all accept this fact when we consider, say, the scientific description of a hurricane: what kind of physical process it is, why it has the properties it has, and under what conditions it may form. But nobody expects that a scientific description of a hurricane will be or cause a hurricane.
>
> Why, then, should we not apply exactly the same standards to consciousness? We should provide an adequate description of what kind of physical process it is, why it has the properties it has, and under what conditions it may occur. As we shall see, there is nothing about consciousness that precludes an adequate scientific description of the particular kind of neural process it corresponds to. What, then, is special about consciousness? What is special about consciousness is its relationship to the scientific observer. Unlike every other object of scientific description, the neural process we are attempting to characterize when we study the neural basis of consciousness actually refers to ourselves – it is ourselves – conscious observers. We cannot therefore tacitly remove ourselves as conscious observers as we do when we investigate other scientific domains.

[Edelman and Tononi, 2000, Pages 12/13]

Rather than untie a 'world knot' I will reveal the morass of systemic gestalts embedded in these statements. To start with *"The answer, we believe, lies in recognizing what scientific explanations in general can and cannot do."* First, the authors are unaware that in the world of science inhabited by the authors there is no explanation (why/causality– that which necessitates that nature unfolds the way it does) and there never was. This was recognised 350 years ago (see philosopher David Hume [Hume and Steinberg, 1993] and the E. Nagel quote in section 2.1) and nothing has changed since. From the lowest level quantum mechanics up, there is no causality delivered by any scientific statement. That is, there is no explanation by science.

We scientists are definitely overdue for some technical specificity in this area. In my mind 'explain' means 'why', not 'what'. I am sure they are not clear on it, when they should be.

Second, where exactly is it that specifies what science can and cannot do? As repeatedly reported here, an explicit body of work that specifies what science does exists nowhere in science. It's not written down. It is a behaviour passed on by imitation. I will actually scientifically measure the behaviour and report it later in this book. Without explicit training, the authors presuppose they have an innate grip on the complete picture of what scientists can or cannot do (what is possible), when in fact all they know is what scientists currently actually do. There can be a major disconnect between these two things and they would not know it.

Now consider *"Scientific explanations can provide the conditions that are necessary and sufficient for a phenomenon to take place, can explain the phenomenon's properties, and can even explain why the phenomenon takes place only under those conditions."* Again: three counts of the wrong usage of the words explanation/explain. The words should not be used in this way. The readership is being misinformed. Next is a problematic use of the word 'phenomenon'. Phenomena are 'the natural world as revealed in contents of a scientist's P-consciousness'. I must repeat: A statement about the apparent sequential orders or concomitance of phenomena does not address causality. I would rewrite the statement as follows: *"Scientific statements can provide the observable conditions that are apparently necessary and sufficient for a phenomenon to take place, can be predictive of the phenomenon's observable properties, but do not account for why the phenomenon takes place only under those conditions."* The problem is again shown as being one in which the authors are clearly unaware that the methods used do not connect with underlying causality. The belief, by the authors, that causality/explanation results from their scientific behaviour, is uninformed by analysis or explicit scientific training.

Next: *"But no scientific description or explanation can substitute for the real thing. We all accept this fact when we consider, say, the scientific description of a hurricane: what kind of physical process it is, why it has the properties it has, and under what conditions it may form. But nobody expects that a scientific description of a hurricane will be or cause a hurricane."* Once again, description (what) is confused with explanation (why). Scientific behaviour is clearly understood by the authors as descriptive (their final chapter's heading acknowledges this reality – see below). But there is no latitude for scientists to impute description as any form of explanation. I run across this confusion continually in the scientific literature, not just here. It doesn't mean science cannot explain. It means that we presently don't do it.

The only context in which the phrase *"... can substitute for the real thing"* makes any sense is if the authors presuppose that a science of consciousness was ever obliged to or has delivered the 'real thing'. Otherwise, why would the statement be made at all? Scientific statements are predictive of appearances. That's it. They predict how the natural world will appear, as contents of a scientist's consciousness, when it is observed by a scientist. In the present tacit paradigm, nothing else can be claimed by science as a result of what we do. Science *never* delivers the 'real thing', so the fact that the authors claim no delivery of the 'real thing' in a consciousness context is simply a trivial statement that does not speak to the process of an account of consciousness. Rather, it speaks to the predisposition of the authors that 'the real thing' might have been an expectation for a science outcome. That is, they consider that, in some way, elsewhere in science that 'the real thing' has been delivered by scientific behaviour, when it never has.

The next paragraph in its entirety represents the most direct touch between the authors' paradigm and my own. It is so close that it's an amazing revelation of the great mental gulf that is involved between us and how it emanates in two directions from the one set of words. Consider *"What is special about consciousness is its relationship to the scientific observer. Unlike every other object of scientific description, the neural process we are attempting to characterize when we study the neural basis of consciousness actually refers to ourselves – it is ourselves – conscious observers. We cannot therefore tacitly remove ourselves as conscious observers as we do when we investigate other scientific domains."* Yes, consciousness 'refers' to us/conscious observers. But *"We cannot therefore tacitly remove ourselves as conscious observers as we do when we investigate other scientific domains."* What? We *never* remove ourselves in the way that is stated. Objectivity *does not do that.* It never did. We have only ever predicted/described how the natural world appears in our contents of consciousness. What we have tacitly been doing is pretending that we did actually perform that miraculous deed of 'removal'. Objectivity merely renders the resultant natural-world-regularities independent of any *particular* observer, not *all observers*, and certainly not independent of observation itself. The 'removal' outcome being assumed by the authors is not what scientific behaviour actually delivers.

To me, the real specialness of a science of consciousness is that its initial job is *entirely and only to account for the scientific observer* – to literally account for how scientists can do science. I see it as its primary function. Once that is done, *then* science directed elsewhere in the natural world takes on a very different complexion. No scientific statement ever produced/predicted, in the history of science, the existence of a scientific observer or a scientist, and especially never predicts that observer to have P-consciousness of the kind we know. That observing entity – us – is presupposed and built into everything we scientists have done for centuries. The instant that the very same activity is directed at any kind of account of how we can do science, then the entire edifice of science is logically compromised in a way that the authors' words seem to acknowledge but actually completely miss. The science of consciousness is not just a 'special relationship with the scientific observer'. Such science *is entirely and only about scientifically accounting for the scientific observer*. And in that circumstance, the possibility of success is intrinsically and fatally compromised.

Therefore, right at the start of the book, the authors are revealed as being unaware of the paradigm they are in. They are unable to articulate the true nature of the problem they face. They do not understand how it is that their chosen paradigmatic approach fails. It fails in exactly this way: unlike any other scientific endeavour, when they have completed their proposed scope of work and done the science, they will be unable to hand over a body of work to an engineer with instructions on how to build an artificial inorganic consciousness. They will hand us a set of correlates, just like every other scientist has ever done. However, in this case, because the correlates have been acquired at an evidence boundary condition, what is handed over is completely empty of what an engineer needs to build an artificial version of the natural original phenomenon.

Take artificial fire or artificial flight as an example. What results when the 'necessary and sufficient conditions' are established artificially, is fire and flight, respectively. The reason we can do this is that the essential (physics) ingredients are what we literally bring to the artificial construction. We can do this because we know how it appears (burning and flying stuff), not because we have accessed the scientific 'real thing'.

In the science of consciousness this is not what is handed over. Instead, engineers will be given a set of abstractions like 'group selections' or 'information integration', and told it's someplace in neurons. Well in exactly what place and in what form is this 'information'? How big is a 'neuronal group' and exactly what constitutes one, so that I might build one? If I compute a model of it, will consciousness still be there? What is the role of consciousness in a 'neuronal group' or in 'information integration'? How much of the biology is absolutely mandatory and how much is not? Exactly what bits of the neurons are essential? Are astrocytes essential? Do we replicate cellular communication literally? Or must we simply communicate between cells by any means? Which bits of the molecular and atomic and genetic and hormonal and energy-producing, regulatory and immune mechanisms and so forth, are not essential and which are?

The ABC-correlate science proposed as their solution is incapable of delivering the essential ingredients of consciousness to an engineer to build because those ingredients *are not actually found by the consciousness science* in the way that they are found elsewhere in science, where at least the actual outward appearance of the underlying causality is directly available. What is actually handed to engineers is not presented in the way needed for it to be built. In the case of flight, for example, the authors would hand over a statement saying *"Flight is highly correlated with the observed startling of observed flocks of birds"* (birds = neuronal groups, for example). The engineer will be given a 'startle instruction' and a picture of 'flocking' (the startle correlates of flocking), and the engineer will be expected to build an aeroplane.

Moving on, the last chapter is entitled 'Prisoners of Description'. Based on what has been discussed here already, you would think that my ideas and the authors' would dovetail. But alas, no. The first interesting difference is that the authors are hostages to the philosophical discourse on XYZisms. On their page 215 is a whole section on 'philosophical claims', in which they go through a raft of XYZisms and declare their selections. The presupposition here is that any of that activity has any bearing on understanding, describing, explaining or building consciousness. The philosophy has no evidentiary basis and explains/predicts nothing testable. I have already shown how philosophy is not even on the field of the game. To act as if it is, is to participate in one of the major routes to failure and part of the syndrome I am trying to reveal. If you think that philosophical positions matter then you have already lost. That is how this appears to me. There is a great gulf here.

Next in their chapter 17 we find:

> Let us consider further the implications of our view of consciousness, particularly for what we have called biologically based epistemology. As we discussed in the first chapters of this book, the study of consciousness as a scientific subject casts a sharp light on a special problem faced by the scientific observer. As long as his description leaves out his phenomenal experience and he can assume that such experience is present in another observer, they can both give a description of the physical world from a "God's eye" view. When the observer turns his attention to the description of consciousness, however, he must face some challenging issues. These issues include the fact that consciousness is embodied uniquely and privately in each individual; that no description, scientific or otherwise, is equivalent to the experience of individual embodiment; that there is no judge deciding categories in nature except for natural selection; and that the external description of information by the observers as a code in the brain leads to paradox. These issues pose a challenging set of problems: how to provide an adequate description of higher brain functions; how information arises in nature; and, finally, how we know – the central concern of epistemology.

> [Edelman and Tononi, 2000, Page 208]

Consider first the words 'biologically based epistemology'. You will find in my later pages an almost radical neuro-biologising of *scientific behaviour* (not an 'explanation of consciousness') along Darwinistic competition/complexity theory lines. That is one difference. Another is that I am mute on whether I speak to an account of epistemology. I need no concept of it to make progress. The only thing I need is for the scientist to make statements (utterances) that are predictive to some useful degree. That is the level at which my proposition operates, and I consider the word epistemology irrelevant. My approach can be implemented without mentioning it.

Consider the next the sentence "*the study of consciousness as a scientific subject casts a sharp light on a special problem faced by the scientific observer"*. See the subtle difference between our understandings of the problem? I consider the scientific study of consciousness to literally *be* a scientific account of the scientific observer. Somehow, because of the authors' life-course in science, they find (1) the concept of a scientific explanation of a scientific observer and (2) the science of consciousness somehow disconnected. This is an example of the paradoxical state of the nascent science of consciousness. The authors cannot see that the explanation of consciousness is about how they can be scientists in the first place. A science of consciousness is, to them, a process of simply putting another advance on the great pile of advances labelled epistemology. I, on the other hand, cannot imagine how they can be thus blinded to their own reality as scientists. The authors' overriding presupposition has the form of an implicit scientific norm: *"I can never explain how it is that I do science, and 'what I do' is the way new scientific knowledge arises"*. Apparently, scientists can never be evidence of anything! We scientists seem to have a very unique position in the natural world – the underlying natural processes – causality – that enables what we do is off-limits, while at the same time we accept, without ever explicitly assessing it, that 'what I do' is all scientific behaviour can be. This is evidence of the tacit taboo (described elsewhere here) operating at the heart of the science.

Next, consider *"As long as his description leaves out his phenomenal experience and he can assume that such experience is present in another observer, they can both give a description of the physical world from a "God's eye" view."* This account of the problem misses the reality that there would be no 'God's-eye view' if there was no P-consciousness. The whole idea of 'leaving it out' is an impossibility. In a previous discussion we saw that this 'view from nowhere' does not exist. It is an agreed projection made possible by the P-consciousness of scientists. The implicit presupposition here is that the 'God's-eye view' has pulled you out of the land of subjectivity into a land of material truth, where all the 'scientific' answers are. My approach recognises that *there is no such thing as the 'God's-eye view'*, and that explaining how we can be tricked into thinking there is one is the actual problem at hand, and that an account of how is part of an explanation of *scientists*.

Consider *"When the observer turns his attention to the description of consciousness, however, he must face some challenging issues. These issues include the fact that consciousness is embodied uniquely and privately in each individual; that no description, scientific or otherwise, is equivalent to the experience of individual embodiment;"* Once again we run into the characterisation of an apparently special situation in the science of consciousness' inability to deliver the 'actual experience', when no science has ever done that. What science does is predict the contents of the observing scientist's consciousness. This is not the process causing the scientist's ability to observe. There is a deep seated, obvious and unacknowledged confusion operating here.

Ultimately, then, the mismatch in thinking is centred on a difference in understanding of what constitutes scientific behaviour, scientific evidence, and the role of the consciousness of scientists in the science of consciousness. The evidence confusion imbues the text. The authors are confusing scientific evidence with the results of objectivity. Scientific behaviour both uses objectivity, and simultaneously evidences its existence/use. To this latter evidence I offer a challenge – try denying it: *"Scientists absolutely must use objectivity or they are not being scientific, and when they do that, there is no evidence of an ability to be scientifically objective or that the objectified was actually 'objectified out of anything'."* Sounds crazy, right? Well that's what we do, methodologically, every day, in the ABC-correlates science of consciousness. The authors are missing the fact of an ability to be objective, as a separate, and just as evidenced, reality for a scientist.

6.2.1 Ok. Enough

The above rant was deliberate. It was designed to reveal the mismatch in thinking as a classic example of dialogue within the cusp of paradigm change, and of the debate between alternatives. It is a topic that Thomas Kuhn covers at length and whose advice I took:

> Briefly put, what the participants in a communication breakdown can do is recognize each other as members of different language communities and then become translators. ... Since translation, if pursued, allows the participants in a communication breakdown to experience vicariously something of the merits and

defects of each other's points of view, it is a potent tool for both persuasion and conversion. But even persuasion need not succeed, and, if it does, it need not be accompanied or followed by conversion. The two experiences are not the same.

[Kuhn and Hacking, 2012, Pages 200/201]

... the choice between competing paradigms regularly raises questions that cannot be resolved by the criteria of normal science. To the extent, as significant as it is incomplete, that two scientific schools disagree about what is a problem and what is a solution, they will inevitably talk through each other when debating the relative merits of their respective paradigms. In the partially circular arguments that regularly result, each paradigm will be shown to satisfy more or less the criteria that it dictates for itself and to fall short of a few of those dictated by its opponent. There are other reasons, too, for the incompleteness of logical contact that consistently characterizes paradigm debates. For example, since no paradigm ever solves all the problems it defines and since no two paradigms leave all the same problems unsolved, paradigm debates always involve the question: Which problems is it more significant to have solved? Like issues of competing standards, that question of values can be answered only in terms of criteria that lie outside of normal science altogether, and it is that recourse to extern criteria that most obviously makes paradigm debates revolutionary.

[Kuhn and Hacking, 2012, Pages 109/110]

This section, then, is my lead-footed attempt to translate from the quoted language of another paradigm to mine and back. Kuhn also predicted long sentences in such translations! With apologies for repetition and wordiness, I hope I have adequately translated across this particular Kuhnian divide. I hope I have directly pointed at the presuppositions within the quoted text, and have translated them into the form in which the issues exist (or not) in my own proposition, which I hold as simpler and less presupposition-laden. My approach certainly needs fewer technical words (I need not discuss or rely on the words 'information' and 'epistemology' or any sort of XYZism, for example). To contrast my approach and theirs is to encounter a truly Kuhnian exemplar of trans-paradigmatic milieu-mangling. My words talk through or across theirs. I am as convinced of my approach as they are of theirs. We are both, therefore, right. However, we are actually being right in two different paradigms. Two different worlds. Mine is the newer, and is a foreign land to them. I know theirs. They will claim mine to be wrong, and may defend their position indefinitely.

That is the process Kuhn revealed happens time and time again in the history of science. I, for one, am actually encouraged by the appearance of such obvious anomaly.

Here's another.

6.3 Manifestations of blockage #2 — engineering

Around 1948, during a lecture by John von Neumann (reportedly at Princeton although the original source is elusive), an interjection arose to the effect that a thinking machine is impossible. Von Neumann has been quoted to have responded thus:

> You insist that there is something a machine cannot do. If you will tell me precisely what it is that a machine cannot do, then I can always make a machine which will do just that.

John von Neumann

If the interjector responded to this by proposing that the apparently 'impossible thing' to demand of a machine is to be a scientist doing science on the radically novel and unknown, von Neumann may have been given pause to think. How can anyone 'tell precisely what it is the machine cannot do' when the machine must do something unknown? If you can do that then it is not unknown! What von Neumann has done is throw up the 'black box' argument.

To engineers, the black box epitomises good design and carries with it all the hallmarks of objectivity. Because of the professional locales I tend to inhabit, more often than not, when the conversation turns to consciousness, the black box is wheeled out as way to render the engineering world invariant to considerations of consciousness. It sounds compelling, and I remember myself thinking that way once. The way black boxes tend to be explained goes something like this (in engineering-speak): There is some system I am interested in capturing in technology. Let's say some bucket chemistry done by hand that I want to automate with a machine. It has inputs. It has outputs. The relationship between the inputs and the outputs is well established. Whether by hand or by robot or by steam or by electronics, I know that if input variables relate to output variables (in the form of a 'mapping' or 'function'), nobody cares how it happens. I can hide the entire affair inside a black box. One implementation and another are, for all practical purposes, indistinguishable.

All of which is perfectly fine reasoning and consistent with hundreds of years of successful practice. Then a question invariably comes about, put forward as some kind of clincher. *"I can do this with absolutely anything, so how can consciousness make that any different? What possible reason must I have to think that consciousness is the difference between a successful black box and one that is not?"*

> In a sense that I am unable to explicate further, the proponents of competing paradigms practice their trades in different worlds ....that is why a law that cannot even be demonstrated to one group of scientists may occasionally seem obvious to another. Equally, it is why, before they can hope to communicate fully, one group or the other group must experience the conversion that we have been calling a paradigm shift. Just because it is a transition between incommensurables, the transition between competing paradigms cannot be made a step at a time, forced by logic and neutral experience. Like the gestalt switch, it must occur all at once (although not necessarily in an instant) or not at all.
>
> Thomas Kuhn [Kuhn and Hacking, 2012, Page 149]

The black box defender has not seen that the thing that I am trying to explain is not a property of the black box. The black box defender has not seen that an account of P-consciousness is unlike any account they have ever attempted. The problem that is being solved is that which enables them to create black boxes in the first place. The black-box defender cannot see that 'inputs' and 'outputs' (I/O) are things that were acquired with their own P-consciousness, and that their P-consciousness has enabled them to separate out all the contextual relations between the inputs, the outputs and the environment within which inputs and outputs are expressed.

I can carefully explain to the black box defender how black boxes don't even address the problem I am asking them to solve. When I do this, I am careful to make sure that I am not painting the black box a broken concept or bad idea. I try to stress how the black box's existence is 100% predicated on the involvement of a P-conscious entity that comprehends that the black box is even there, and that it carries out the functionality assigned to it. I have asked them to solve a problem in which the space of available solutions does not include a black box for reasons that cannot be seen by the black box defender. To the black box defender I speak in tongues. I ask them to consider *"What it is like to be a black box?"* and why the answer is important. I can point out the black box defender's own P-consciousness, and how it was involved in the creation of the black box. They will be unable to see why any of that matters.

There is even one logical sweet spot that is absolutely game-over for the black box defender. It is the von Neumann retort. Ask them *"What happens if the black box functionality is to be a scientist?"* This is a circumstance where I/O (input/output) is unknown *by definition*. A scientist cannot be a scientist unless the I/O is unknown. You don't know what the natural world is doing, and you want the black box to create a scientific account that is also unknown. So you can't design the black box. In the face of this, the black box defender will tell you about machine learning and genetic algorithms and the like, as if their operation is in some way identified with that of a scientist. The inability to see the problem space is right there. The black box defender will be unable to see that such an approach to a black-box-scientist will remap the natural world into categories its designer built into the black box in all the learning algorithms. The black box has been told, by a P-conscious human, everything it needs to know about novelty and 'what it looks like'.

Can you see the weirdness of this? There is a rock-solid *definition of novelty*. In other words, there is effectively nothing unknown at all. There is a sense in which you already know it or you wouldn't be able to detect it. The adage *"To those who only have a hammer, all the world's problems look like nails"*, in this case, is literally applied to the black box by its designer. The 'black box scientist' encounters the natural world depicted by what the designer has told it, not the actual natural world. The black box's account of the natural world can depart that of the designer in a fundamental way, and the black box will have no clue. Even in the face of such brute logic, the black box defender will be unable to see this. This is the bind so well captured by Kuhn.

The above novelty argument leads to an interesting end-point for a black box defender's machine learning solution. Faced with radical novelty, the 'black box scientist' makes random guesses and self-applies black box ideas to its own job as a scientist. If there is no information supplied by the designer to handle novelty, the black-box can simply randomly generate outputs, test them against inputs, and then, when some level of probability is reached that the inputs match the output in some regular way captured by the black box, the 'black box scientist' decides it is done. There is some kind of empirical exploration in this. For those in the field, Bayesian (probabilistic) learning is a common tool used for this purpose. So in answer to the novelty argument, the black box defender is quite happy for the 'black box scientist' to be totally unlike any human scientist. The result has the exploratory blindness of evolution by natural selection, and may even be successful in some sense, but to a human scientist observing the 'black box scientist', the black box lives somewhere that bears little resemblance to the natural world of the human scientist.

The 'black box scientist' will hand over strings of gobbledygook that make its inputs (masses of numbers as the data from the best video camera ever) and outputs (masses of numbers that control myriad probes the black box uses to explore its environment) match really well, but that have an impenetrable relationship with the natural world that the human scientist sees. Despite all this logic, the black box defender's position remains. There is no place for consciousness in this, despite consciousness having been used to mediate the entire thought process.

Now consider a further subtlety in this. We have just examined the fact that the black box 'sees' the world as a bunch of numbers that might come from an elaborate 3D video camera that can act like an electron microscope or a massive radio telescope. The black box might be able to manipulate atoms and move mountains. Its interface to these controls is also masses of numbers. The black box defender cannot impute any experience (P-consciousness) in these flows of numbers. That would be to invoke the presence of something the black box defender is denying the relevance of. That being the case, how can the black box distinguish between itself and the natural world it is scientifically encountering? Paradoxically, the one thing that P-consciousness provides, an ability to distinguish between self and not-self, is missing from the black box. That is, the black box has no objectivity – something provided in humans as a consequence of our subjectivity. "*But then,*" the black-box defender continues, *"we can put a model of the physicality of the 'black box scientist', and all its I/O, inside the black box, thereby enabling it to distinguish between itself and what it is attending to."* Again this sounds plausible, but in fact merely compounds all the problems the black box has in being a scientist, because the black box has to interact with the natural world in novel ways, and even build novel instruments (add to its own I/O), something for which no human can prepare it, if the black box is truly to be an authentic scientist. And so the argument goes.

The mechanism of the Kuhn quote is operating here in the black box defender. I have logic, but because the logic involves a mental view of scientists that the recipient of my logic cannot see, that logic will fall on deaf ears. This has happened to me time and time again over a decade. What do you do about this? One thing to do is write a book about it and hope it will help the gestalt switch do its work. Centuries of ingrained habit will not go quickly or quietly.

6.4 Summary

This chapter extends Kuhn's work on the psychology of scientific change behaviour to examine its implications for the change to science proposed here. It explored more deeply the notion that scientific behaviour, as currently practiced, is an entirely neurological phenomenon. Kuhn originally revealed how the 'received view' is more than just a collection of theories. The transition to the new paradigm is more than just the installation of new technical insights. Scientists literally see the world differently in different paradigms.

To illustrate, Kuhn's own account was analysed, an example from the recent consciousness literature was presented, and my personal experience in engineering was then related. These were used to tease out the fundamental incompatibilities at play here.

Scientists of the current era might think we know what science can/cannot do, when I claim all we actually know is what we currently do. Scientists of the current era might think a science of consciousness is like any other science campaign, but I claim it is actually the science of the scientific observer, acting primarily in an explanation of how scientists can operate at all, and demands something extra – over and above expectations anywhere else in science – an account of a first-person-perspective (1PP). Scientists of the current era might think scientific evidence is an act of objectification, when I claim that it also implicitly evidences the P-consciousness that is scientific observation. Scientists of the current era might think they have a formally established understanding of the difference between description and explanation, when I claim that the two are routinely used inconsistently throughout science, and that where formal explanation (as defined here) might be thought impossible, is actually possible and unexamined to that end.

History, Kuhn advises, predicts enormous mental inertia when scientific change seems needed. Great battles will surround any apparent claim of crisis and the need for paradigm shift. This chapter spelled out what such conflict might look like for those involved. The process of change is more than just a swap-out of old with new scientific statements. It requires the reeducation and establishment of a whole new scientific world-view. This leads us back to the centrality of neuroscience in scientific behaviour: the neurodynamics of 'scientific statement construction' are at the centre of it all, and will be addressed in more detail in a later chapter.

Chapter 7

Cultural Learning Theory for Scientists

This chapter takes a topic to appreciable technical depth in order to more formally prove how scientists can operate happily without any formal science of scientific behaviour. For those readers inclined to avoid details like this, I recommend you get a cup of tea, squint a bit, then push through and you'll pick up the gist well enough. Take heart – most scientists are in the same boat. Alternatively you may skip to the summary at the end. In this chapter, if I use the word 'knowledge' it can be replaced by the technical sense of 'statement' used elsewhere here. The word knowledge seems more apt in the context of the chapter.

7.1 Introduction

In the last couple of decades, behavioural scientists grappling with social learning have converged on a reasonably stable and useful learning framework based on an intersubjectivity-mediated behaviour called 'perspective taking'. It has been used to compare and contrast human and non-human social learning, changes in learning during organism development, and to take an evolutionary perspective on learning capabilities in humans and non-humans. The framework came about when trying to understand human culture transmission and whether group-based social knowledge transmission in non-human species is 'culture' or something else [Tomasello *et al.*, 1993].

This work upgrades the above learning-theoretic model for the purposes of examining scientific behaviour. Using well established empirical neuroscience and cognitive science, the intersubjectivity model for learning extends relatively easily to include possibly non-intersubjective sources of knowledge such as scientific behaviour and rote-learning. Scientific behaviour involves an encounter with natural-world novelty for the purposes of creating new knowledge. This is quite distinct from the novelty encountered by a novice who is merely ignorant of existing knowledge. Scientists start as novices in respect of new (or new incremental) knowledge. At the end of the science process, the scientists involved become the first masters in the subject area, and the new explicit knowledge appears in the scientific literature.

With the learning framework upgraded and sufficiently expressive, a learning-theoretic treatment of scientific behaviour is to be undertaken. The first outcome recognises this chapter as an example of the process that it seeks to document, so the activity is intrinsically self-evidencing and is shown to be self-consistent under the revised model.

The second outcome formally reveals science's self-governance vacuum. Science self-governance is explicitly demonstrated to be non-existent because scientific behaviour is passed on through imitative learning only. Imitation creates very successful self-regulation that faithfully maintains scientific behaviour. But because the behaviour itself is formally undocumented and not subject to review, there is no explicitly managed way the behaviour can adapt to defend against novel circumstances or requirements that are beyond the reach of the (imitatively maintained) existing behaviour. Difficulties in the scientific treatment of consciousness are shown to be that type of challenge, and have only been presenting that way for the last twenty years or so – the duration of the mainstream empirical scientific study of consciousness. The zero self-governance state of science is not in place because of choice by any living scientist, and the lack of management of it means that no scientist can access any principled argument resulting in the explicit, documented choice to scientifically behave in any particular way, whether challenged by the science of consciousness or not.

7.2   The existing intersubjectivity-based learning framework

In the original article behind this chapter and its associated commentary, the cultural learning categories can be understood by noting that social learning arises as a result of (1) intersubjective perspective-taking and (2) the active and passive status of the interactants as a determinant of perspective. The learning categories have an interesting structure when organised graphically as shown in Figure 7.1. The examined learning categories are [Rogoff *et al.*, 1993; Tomasello *et al.*, 1993]:

(i)   *Imitative learning*: The learner internalises something of the demonstrator's behavioural strategies. This may occur either inside or outside a pedagogical context. The learner is active and the social world passive.

(ii)   *Instructive learning*: The learner internalises the instructions of the teacher and uses them subsequently to self-regulate their own attentional, mnemonic, or other cognitive functions. The social world is active and the learner is passive.

(iii)   *Collaborative Learning*: Takes place when neither interactant is an authority or expert; intersubjectivity is symmetrical and dynamically active/passive. Interactant roles are still separate efforts to take the perspective of the other. Joint knowledge constructions are later individually internalised.

It is relatively straightforward to run the Figure 7.1 categories with test cases. Consider the master/novice relationship in a context of, say, hunting or swimming. The instructor behaves, and the learner imitates behaviour with a view to achieving the same outcome (hunting and swimming, respectively). A mental model for the behaviour is constructed by the novice. Next, consider two novices practicing a new dance or two children playing with a new toy robot. They are engaged in collaborative learning. Imitative learning can result from playing tennis with the coach. Two tennis novices playing each other results in collaborative learning to the same or a similar effect. In contrast, if the tennis coach instructed the rules of tennis, the novice would learn an abstract model for tennis, not tennis playing behaviour. If two novice tennis players taught each other the rules of tennis, then collaborative learning of an explicit model has occurred.

7.3 Attributes of a more general model

The process of upgrading and generalising the learning model was started by simply asking about the nature of the empty boxes in Figure 7.1, and then associating all the boxes with neuroscience knowledge.

To start the analysis, consider that an international body is responsible for the maintenance of the rules of tennis. Governance of the rules of tennis involves their review and update based on their suitability in novel circumstances. The rules are externalised explicit knowledge existing in, say, a book. A novice might sit with this book of rules of tennis and learn tennis rules, resulting in internalisation of an explicit model for tennis, yet never play tennis. Notice that there is no imitation and there has been no instruction, yet there has been learning of an explicit model. This kind of learning of an externalised explicit model is what we call book or rote learning. It happens without intersubjectivity. In general, Figure 7.1 needs to account for learning from externalised sources with and without intersubjectivity.

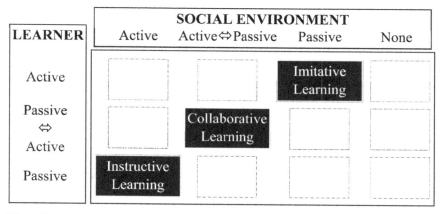

Figure 7.1.   Table of intersubjectivity-mediated categories and learning kinds as per [Tomasello *et al.*, 1993]. The empty boxes are populated by considering learning modes other than purely intersubjective.

Figure 7.1 does not explicitly distinguish between 'abstract-symbol-reportable' knowledge (such as the formal rules of tennis) and 'non-abstract, behaviour-only-reportable' knowledge (tennis playing behaviour according to the rules). These are called 'explicit' and 'implicit' models, respectively, and are more formally defined below. An upgraded Figure 7.1 needs to include this distinction if it is to concisely account for scientific behaviour, which involves both kinds.

Figure 7.1 presupposes existing static knowledge that is then learned by the novice. It does not include a learning class where the knowledge is impossible to learn or teach because it does not exist yet. There is no master and there is nothing to imitate or instruct. Everyone is a novice. The natural world itself is the instructive master. More generally, Figure 7.1 must be upgraded to recognise knowledge change dynamics, information flow during knowledge creation, and how this determines the behaviour of all the entities involved. In particular, science creates and manages externalised explicit models (set *T*). Figure 7.1 needs to include the creation of novel knowledge in the form of implicit internalised models, explicit internalised models, and externalised explicit models, created with/without intersubjectivity.

Figure 7.1 does not explicitly recognise the embodiment and embeddedness of learning contexts. Figure 7.1 situates learning as occurring in the presence of another cognitive agent (a social environment) in the three senses shown. In reality all interactants, as cognitive agents, are embodied and embedded in a shared natural environment [Robbins and Aydede, 2009; Varela *et al.*, 1991]. Regardless of a learned model's status as implicit/explicit, Figure 7.1 needs to be upgraded to include embodiment and embeddedness if it is expected to account for scientific learning, which is a specialised case of learning an explicit model with/without intersubjectivity. To do science is to form an account of the natural environment, from within that natural environment, using the cognitive faculties of an embodied agent. So far, the only empirically observable entity capable of scientific behaviour as defined here is the human, embodied and embedded as discussed, with a biological substrate for its intelligence [Carruthers *et al.*, 2002; Kitcher, 1993].

Figure 7.1 shows no explicit recognition of a learner's internal experiences (first person perspective, P-consciousness as defined in Chapter 3). To make the role of phenomenal consciousness explicit in a model of learning, a Figure 7.1 upgrade would, at a minimum, divide P-consciousness into two kinds: (1) exogenous, called sensory/perceptual P-consciousness, resulting from peripheral nervous system information and, (2) endogenous P-consciousness, which includes the imagined and dreamt counterparts of (1) and other kinds such as emotions.

Figure 7.1 also does not explicitly account for imagistic learning. In imagistic learning, instead of a source outside the learner, the learner imagines (using P-consciousness) some invented knowledge or state of affairs based on reflective endogenous P-consciousness (internal imagery). In the Figure 7.1 context, a useful way of thinking of this kind of learning is that the novice replaces the direct experience of a peer or master with endogenously generated imagery appropriate to some level of functional equivalence of a peer or master. For example, collaborative learning is a context for imagistic learning, where the intermittently active learner imagines they are the master, for the purposes of collaborative learning. Imagistic learning includes, amongst others, diagrammatic/visual and sentential/auditory, used separately and cross-modally [Magnani, 2009].

Recognising embodiment and embeddedness explicitly reveals a subtle aspect to understanding learning through intersubjectivity. Consider master => novice verbal instructive knowledge K1 = "*X is true*". Now consider exactly the same the same explicit knowledge communicated verbally by consistently screaming it out, for example, K1 = "*X IS TRUE!!!*". The same explicit knowledge has been transferred, but the novice has actually experienced two different behaviours and received two different lessons. The first is the explicit K1= "*X is true*", and the second (amongst many possibilities) is the implicitly evidenced K2 = "*Teaching people involves screaming the knowledge*". Different implicit knowledge K2 was transmitted in the first case by virtue of K1 not being shouted. Every intersubjective interaction transmits both kinds of information. This is intrinsic and unavoidable through the mere presence of embodied intersubjectivity; the two kinds of information are inextricably entwined.

There is another subtlety to these aspects of intersubjectivity. An instructor delivering K1 = "*X is true*" also delivers knowledge of the form K2 = "*This instructor tells me things*". Paradoxically, K2 is logically more certain than K1. This is a self-evident general property of intersubjectivity upon which science ultimately depends. In the example, the novice has direct evidence of the claim K2, but only hearsay evidence of the truth of K1. Learning through instruction relies on the authority of the instructor as master in a master/novice interaction. This is important in an account of scientific learning (new knowledge creation).

In creating new knowledge, science only accepts evidence in the direct form of an experience of (the unique logical consequences of) X in the natural world. Only then is the claim K1 eligible to become scientific knowledge. In science, a master is accepted as an authority on the truth of X by virtue of knowledge of X having arisen in this way, not because the master uttered K1. The double-delivery nature of intersubjectivity means that whether an interactant is master or novice or peer, during active (instructive learning) or passive (imitative learning) or active/passive (collaborative learning), a cognitive agent cannot avoid simultaneous delivery of both the explicit and implicit communications discussed above. Figure 7.1 does not recognise this expressive duality.

In the original paper, there was the idea of mimicry, where a behaviour X is imitated for reasons other than to be an instance of role X [Tomasello *et al.*, 1993]. This is called acting, and is a useful way to see nuances in the limits of Figure 7.1. Regardless of the motive for it, acting is an interesting case where the outward signs of a role X are present, but the internal models driving the behaviour are not necessarily the result of any actual expertise in role X. An actor-astronaut in a movie is not an astronaut. In general, acting a role X is not an instance of role X and is not an instance of learning X. Yet acting a role X can play a part in teaching role X. Teaching X can be an effective way of learning X. If an actor exhibited instructive behaviour identical to a master, possibly the same or similar level of instructive learning (in a taught novice) may result.

A video presentation of an instructor has an identical effect in terms of the learned explicit model, but a quite different implicit learning outcome (K2 in the examples above). The video media is, in effect, an actor. In and of itself, it knows nothing of the instructed knowledge. In either case, the novice may be unaware of the acting, yet just as informed. The same goes for imitative learning. Learning teaching by teaching also has this aspect of acting to it. Prior to mastery, the novice teacher is, to some extent, acting like a master of teaching in order to learn how to teach. Acting like a (instructive/imitative) teacher and teaching someone to act like a (instructive/imitative) teacher both fit differently into the scope of an upgraded Figure 7.1. Currently these are not obvious in Figure 7.1.

7.4 Definitions: Implicit, explicit and reportable models

For the purposes of upgrading Figure 7.1, the working definition of the terms implicit model and explicit model are:

Implicit Model refers to that portion of a configuration of the memory of a cognitive agent (knowledge) that is only reportable by means other than abstract symbols. E.g. an ability to walk. This has also been extensively called 'tacit' knowledge in the literature.[a]

Explicit Model refers to that portion of a configuration of the memory of a cognitive agent (knowledge) that is reportable by means of abstract symbols, through the existence, *a priori*, of other implicit models that act in support of abstract symbolic communication (e.g. spoken language skills).

Reportable means that the external boundary of a cognitive agent may be operated to communicate or interface with the rest of the natural world. Dancing, speaking a language or sending smoke signals are all 'reports'. An implicit model in cognitive agent H1 may be an explicit model in cognitive agent H2, merely by virtue of its reportability as abstract symbols. Such reportability implicitly and automatically creates

---

[a] Michael Polanyi's tacit knowledge is an identity with the implicit model [Polanyi, 1967].

a new class of instructive learning and a new class of collaborative learning. Reportability in abstract symbols automatically makes science possible, because it is directed at the creation of externalised models, which ultimately demands explicit models within the scientist.

7.5 A more generalised learning framework

The Figure 7.1 shortfalls identified in the previous section are

(1) Identification of the kinds of learned models (explicit vs. implicit).
(2) Learning without intersubjectivity.
(3) Embodiment and embeddness of the cognitive agent in all learning contexts.
(4) Knowledge dynamics and information flow for novel explicit/implicit models.
(5) The critical dependence of learning on P-consciousness.
(6) Imagistic learning.
(7) The dual-aspect of intersubjectivity means a cognitive agent cannot avoid simultaneous delivery of explicit and implicit messages.
(8) Externalised explicit models.
(9) Acting in general, and in particular, acting the role of master (instructive or imitative learning) or a peer (collaborative learning) must be able to be distinguished from learning to be a master (instructive or imitative learning) or a peer (collaborative learning). Both of these must be distinguishable, within the model, from using the role of master of X as a means to learn X.

The upgrade is relatively straightforward, and is shown in Figure 7.2. Figure 7.2 accounts for (1) by incorporating explicit (i) and implicit (h) models within the declarative and procedural memory types. These are memory types with a long history within empirical neuroscience [Kandel and Schwartz, 2000]. The learned explicit (i) and implicit (h) models fit nicely into these known kinds of memories. Figure 7.2(p) accounts for (8), externalised explicit models, where the externalised model (say, a scientific journal) is accepted as part of the natural world of the learner. Figure 7.2(d) shows it as a source of non-intersubjective rote learning.

To account for (3), Figure 7.2 recognises the separate embodiment of the novice, master and peer (as cognitive agents), their embeddedness in the natural world and their inherited relationship with each other, the rest of the natural world and the objects in it. Each cognitive agent has the internal machinery of cognition (a peripheral and central nervous system complete with subjective experiences and memory), as shown in the legend. Figure 7.2(b)/(c) recognises item (7) in cognitive agency (m) that in (b) actively delivers explicit messages via (f) to the learner to reinforce explicit or implicit learned models (instructive), and simultaneously via (c), passively delivers implicit messages, also via (f), to reinforce the same models (e.g. imitative learning). Figure 7.2(f) and (g) account for omission (5), the role of P-consciousness. This too has a couple of decades of mainstream empirical science, and one of the few well established properties of P-consciousness is that the physics that creates it is a cranial central nervous system property, and is therefore intimately cohabiting with the above two memory types. As a result of accounting for (5), omission (6), imagistic learning, is automatically covered. What can be found in the literature as 'sensory/perceptual' learning is also present as (f).

Figure 7.2 addresses omission (2) by accounting for non-intersubjectivity based learning from exogenous sources (d), (e) and endogenous source (g). Figure 7.2 addresses omission (9) operationally by populating Figure 7.2 with appropriately behaving cognitive entities. For example, a learner may be placed in the role of instructive master. An actor may be inserted in the role of imitative master. An actor may be pretending to do rote learning. An actor rehearsing a role of playing the part of someone rote-learning could be types (a), (b), (c), but is not (d). Item (9) examples are covered in Tables 1...3. No extra structural features need be included in Figure 7.2 to account for (9). The ability for Figure 7.2 to express these kinds of learning arises from the explicit embodiment and embeddedness of all the cognitive agents in Figure 7.2.

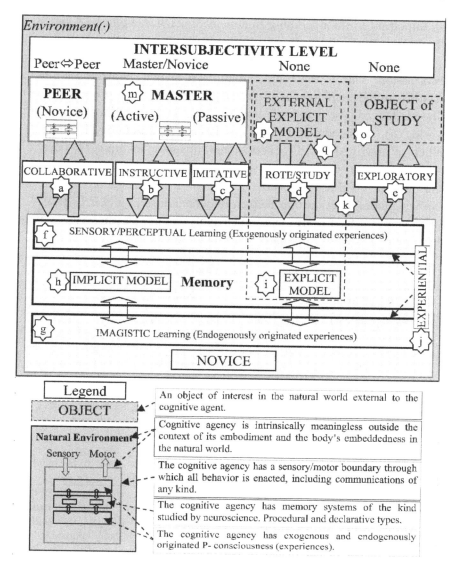

Figure 7.2.   (a)...(e) Learning kinds related to the dynamical learning of (h) implicit models and (i) explicit models. (j) Experiential learning is the collective effect of (f) sensory/perceptual and (g) imagistic learning. (k) Dashed box shows a process unique to humans. (o) is any natural world subset that is the subject of exploratory learning, including scientific learning. (p) refers to the external embodiment of an explicit model via mechanism (q), such as a book or journal.

Important in the upcoming treatment of science, Figure 7.1 omission (4), knowledge creation dynamics, is addressed in Figure 7.2(p), (o) and (g), where the natural or imagined world can act as a source of information. This is the route through which new knowledge creation occurs, through experiences of novelty involving the *a priori* unknown. Knowledge can be explicitly externalised via (q). When the natural world, (o), is encountered via (f), it is called exploratory learning (e). When an imagined world is encountered, imagistic learning, (g) occurs.

Together (f) and (g) may sometimes be found called experiential learning (j). To illustrate, Table 7.1 shows a few cases of implicit and explicit model creation that help in understanding Figure 7.2. Table 7.2 and Table 7.3 also more comprehensively enumerate the various ways that Figure 7.2 deals with many learning circumstances.

Table 7.1. Figure 7.2 model learning examples.

#	Activity/Circumstance	Figure 7.2 context
1	Proto-human H1, without language, has knowledge of what is edible and what is not.	An implicit diet model **h** exists in H1. The implicit model can be instructed by gesture and other non-abstract means via Figure 7.2 **m-b-f-h**. The diet can be also learned from H1 by imitation of eating habits, which duplicates the implicit list **h** in the novice via Figure 7.2 **m-c-f-h**.
2	Proto-human, H1, alone and without language, encounters a novel food and explores its edibility.	An implicit model **h** in H1 has been modified by exposure to the natural world via Figure 7.2 **o-e-f-h**.
3	Two cubs play-act a kill with a stand-in prey.	The cubs learn an implicit model **h** for hunting through peer collaboration. Figure 7.2 **a-j-h**.
4	Human H2 has spoken language skills but the same diet as H1.	The #1 diet model exists as an explicit model **i** in H2. However, extra memory configuration exists, as an *a-priori* implicit model **h** (language skills), that permits the list to be verbally enumerated (a behaviour involving vocalisations). H2 can instruct, as master **m**, the explicit model **i** to conspecifics as per Figure 7.2 **m-b-f-i**.
5	Human H2 carves the diet H1 **h** onto a rock wall as pictograms.	H2 has externalised an explicit model **i** using imaginative learning via Figure 7.2 **i-j-q-p**. The externalised model **p** then exists as an object in **o** for others to encounter.
6	Proto-human, H1, alone	H1 has invented technology via route **o-e-j**, and

	and without language, discovers that a bone can smash other bones, and uses it to hunt.	has an implicit model **h** about its use. This behaviour can be learned by conspecifics as per #1.
7	#4 human H2 learns to act like an astronaut for a movie.	H2 speaks a rote-learnt part, an explicit model (the script) learned via **p-d-f-i**. H2 also has an implicit model for astronaut behaviour also learned from the script via **p-d-f-h**, and via instruction from a director via **m-b-f-h**, and through imitation of video of a professional astronaut at work via **m-c-f-h**. To the extent that **h** can be reported, H2 also has an explicit model **i** that was acquired simultaneously from the two different masters. Rehearsals are learning. Acting like an astronaut, for H2, is not an act of learning to be an astronaut. Acting like an astronaut consolidates the actor's general implicit acting skills.
8	#4 human H2 reads a science fiction novel.	H2 has an implicit model **h** of the reading experience and the physical novel itself (**p-d-j**), which may be used to recall, via **g**, an explicit model **i**, which is evident insofar as the story can be communicated as a master/instructor via **m-b-f**.
9	#4 human H2 goes to a theatrical performance	H2 has an implicit model **h** of the performance and its physical environment (**o-e-j**), which may be used to recall, via **g**, an explicit model **i**, which is evident insofar as the performance can be communicated as a master/instructor via **m-b-j**.
10	A novice scientist trains novice mice to respond to noises and lights.	The mice learn an implicit behaviour via **o-e-j** (Classical Conditioning). At the same time, the novice scientist learns how to train mice to respond to noises and lights by acting as a master, via (**m-b-hi**) *in the novice-scientist in position **m** in Figure 7.2)*. The act of training novice mice is also learning for the scientist.

Table 7.2. Figure 7.2 INTERSUBJECTIVE learning cases.

INTERSUBJECTIVE LEARNING	
EXISTING KNOWLEDGE (p,m)	
Learning outcome: Explicit model (i)	Learning outcome: Implicit Model (h)
(i) Instruction by master (m-b-j-i) The master's explicit knowledge, say, m_i, of some existing external explicit model **p**, is acquired by the learner as **i**.	**(ii) Instruction by master (m-b-j-h)** Master's internal models m_h, and/or m_i of desired learner behaviour is acquired by the learner as implicit model **h**.

	Master achieves this through behaviour other than the learning target.
• Lectures in a profession (e.g. mathematics, law, engineering, language, acting and so forth). • Lectures in the rules of a game. • Lectures in formal teaching rules that underpin a role as master, m, in a learning context. • Elevator/sales pitch for a product. • Lectures in rules of governance and procedural rules of an organisation. • In delivery of any of the above: Use of computer, video or other media based training in support of an explicit model, where the media acts as a master m	• Dog training. • Fledgling pushed out of nest. • Sports coaching. • Teaching how to be a master, m, in a learning context. • Demonstration to teach a corporate procedure. • Car driving instruction. • Baby encouraged to walk.
(iii) Imitation of master (m-c-j-i) Learner acquires explicit knowledge **i** of externalised explicit model **p** through imitation of master's instructive behaviour while instructing **p**. • Learning a formal subject by teaching it. Includes the formal model for the teaching of formal models (i), (iii) and (v).	**(iv) Imitation of master (m-c-j-h)** Learner acquires implicit model **h**, in support of desired behaviour through imitation of master's behaviour. • Cultural norms, etiquette. • Professional/Hobby/personal skill performance. • Dance/sports participation. • Driving (machinery operation) lesson. • Learning to write by writing. • Performance of tribal ritual/rites of passage. • Imitation of behaviours learned from media (TV, computer). • Student teacher gives classes as a means to learn to teach. • Learning how to act like someone involved in or delivering intersubjective learning (i)...(viii) or non-intersubjective learning (i)...(iv). The targeted implicit model covers only the delivery of the behaviour component (going through the motions), not the skill outcome related to the behaviour. For example, learning to act like someone learning to speak a

	language without having any knowledge of the language. • Video or other media used in delivery of any of the above.
(v) Peer collaborative (a/j/i) Learner internalises to **i** the explicit model **p** through experiential **j** access to a virtual instructive **b** or imitative **c** master. Conspecific peer novice(s) enact the corresponding imitative or instructive role as required.	**(vi) Peer collaborative (b/j/h)** Learner internalises implicit model **h** of a target skill through experiential **j** access to a virtual instructive **b** or imitative **c** master. Conspecific peer novice(s) enact the corresponding imitative or instructive role as required.
• Study-buddy homework for a formal subject. • Organisation: Procedural training workshop. • Show and tell of a formal topic. • Actors learning their words from a script.	• Juvenile mammal play. • Acting rehearsal. • Imaginative play/Role-playing/Theatre sports. • Show and tell. • Language acquisition through speech with peer novice. • Fire drill.
INTERSUBJECTIVE LEARNING	
KNOWLEDGE CREATION (p,h,i)	
(vii) Collaborative/Experiential **(o-e,a)-(j-q-p)** Learner uses **j** in the context of interaction with external world **e/o** and/or peer(s) **a** to fabricate novel explicit model **i**, then externalises it to via **q** to **p**.	**(viii) Collaborative/Experiential** **(o-e,a)-(j-h)** Learner uses **j** in the context of interaction with (1) external world **e/o** and/or (2) peer(s) a to create a novel implicit model **h**.
• Scientific behaviour by a group (empirical or theoretical). Including learning theory of the Figure 7.2 type. • Brain-storming. • Creation of new laws by a group. • Creation of literature by joint authors.	• Joint exploration of novel environment. • Joint creation of a new non-written communications method, say a spoken language.

Table 7.3. Figure 7.2 NON-INTERSUBJECTIVE learning cases.

NON-INTERSUBJECTIVE LEARNING	
EXISTING KNOWLEDGE (p,o)	
Learning outcome: Explicit model i	**Learning outcome: Implicit Model h**
(i) Rote/Study (p-d-j-i) Unsupervised learner uses experiential learning **j** in the context of interaction with externalised explicit model **p** to	**(ii) Rote/Study (p-d-j-i)** Solo learner uses experiential learning **j** in the context of interaction with externalised explicit model **p** to

internalise explicit model **i**, after which there is some mastery of the knowledge **p**. • Learning the words of an acting part. • Reviewing a scientific paper. • Study aimed at professional knowledge acquisition such as mathematics or physics or law. • Computer mediated training such as online help, Wikipedia, scientific journal system, computer based training in fire regulations or emergency procedures. • Learning of datasets and procedures. E.g driving rules of the road or Scandinavian monarchs. • Study of organisational procedures and procedures for changing procedures. • Study of rules of governance of organisations and rules for changes in governance.	internalise explicit model **i**, also acquires some explicit/implicit learning **h** about the process of rote learning. This may be the actual goal of the rote study. It may be used to teach someone how to rote-study. Incidental side effect-learning of all of the categories in (i) • Reading to improve (i) rote learning abilities.

NON-INTERSUBJECTIVE LEARNING
NEW KNOWLEDGE (p)

(iii) Exploratory/Experiential (o-e-j-i-q-p) Unsupervised learner uses **j** in the context of interaction with external world **e/o** to fabricate novel explicit model **i**, then externalise it via **q** to **p**. • Empirical scientific behaviour. • Theoretical science. • Creation of literature by single authors. • Creation of new legal statutes. • Fine art creation. • Writing a critical review of a scientific paper.	(iv) Exploratory/Experiential (o/e/j/h) Unsupervised learner uses **j** in the context of interaction with external world **e/o** to fabricate/improve implicit model **h**, then uses the model to operate in that natural environment. • Exploration of novel environment. • Classical conditioning by environmental stimulus. • Imaginative play. • Experimental invention (the wheel)/domestication (fire). • Empirical science method learning. • Learning to discriminate more tastes smells or colours. • Attending an acting performance. Reading a novel (fiction or biography). Machinery operation by operating machinery without supervision.

	• Sports practise/Fine art practise. • Imaginative play that creates/learns a new sport or performance (solo rehearsals of various kinds)

7.6 The upgraded model: Discussion

Figure 7.2 is a novel figure, but the components of Figure 7.2 are all well trodden, empirically supported, existing neuroscience knowledge. As a novel systematisation of existing knowledge, the value of Figure 7.2 is in its role in teasing out aspects of learning within a context of interest. One can look at a particular learning context, say X, and then visit the figure and the tables to find out the way that X has been implemented as an instance of learning dynamics. One might analyze religion or darts or CERN supercollider science or a kitten's first encounter with ice cubes in its water. However, it is the application of the model to science itself that is most relevant here. That application will shortly reveal the subtleties in scientific behaviour relevant to the wider context of this book.

In the confusingly self-referential way that this subject seems to engender, this chapter (and therefore Figure 7.2), is actually an instance of an externalised explicit model (Figure 7.2(p)), and its production is an instance of scientific behaviour identifiable within Figure 7.2. The self-referential, self-evidencing and self-consistent nature of the model is thereby demonstrated. This aspect is a necessary feature of any cogent scientific account of scientific behaviour. It does not mean that the Figure 7.2 model is perfectly accurate or complete, but merely that it has one of the necessary properties of a complete model.

Communication is an interesting way to test the scope of Figure 7.2. The dashed box (k) in Figure 7.2 recognises that human-level cognitive agents are, currently, the only cognitive agents that construct and maintain qualitatively novel explicit models (i) and externalised explicit models (p) using abstract symbol systems that both refer to the natural world elsewhere and that exist independently of any particular agent. Some animals can be taught novel abstract symbolic communications, can demonstrate novel usage of it and possibly even extend it. But these incremental activities are not human-level creation of novelty. No member of the non-human animal kingdom has this capacity. No animals do science. Animals can solve problems and exhibit subsets of (a) collaborative, (b) instructive and (c) imitative learning. However, these are restricted to implicit models (h) as per Table 7.2 and Table 7.3.

Figure 7.2 also provides a clear mechanism by which human and non-human language may arise and be learned without the need for any existing master or prior externalisation of a model, thereby making it consistent with an evolutionary account of the emergence of science. Figure 7.2 also allows for non-linguistic forms of communications, such as mating rituals, music or dance, regardless of the levels of abstract symbolism. Figure 7.2 allows for the scientific study of language (linguistics), the formal teaching of language, an actor pretending to learn a language that is not actually understood, and a novice self-teaching a language through teaching it to other novices. Figure 7.2 is adequately expressive in regards to communications.

Notice that Figure 7.2 has no special explicit mention of scientific behaviour. This is because in reality, scientific behaviour is merely a specialised version of everyday problem solving behaviours [Einstein, 1950; Popper, 1999]. What makes a behaviour 'scientific' behaviour is a result of processes captured in Figure 7.2 that are examined in the next section. Figure 7.2 also accounts for experimental/empirical learning, where the novice or the novice's proxy manipulates the external environment (the object of study in (o)) with a view to experiencing the logical consequences of it.

Table 7.2 and Table 7.3 exist to enumerate instances covered by Figure 7.2. In addition to covering the omission list (1) to (8), the tables have been constructed to illustrate the omission (9) nesting properties of learning to learn to learn..., teaching to learn to teach to ...and so forth. The tables are not exhaustive, and others may extend and explore them. A key feature of the tables is that P-consciousness (f) and (g), as represented by (j), appears in every single panel of Table 7.2 and Table 7.3. Without it, learning of the human kind simply ceases. The fact that (j) has no scientific explanation is irrelevant. The fact that one might imagine learning without the existence of (j) is likewise irrelevant because we have no empirical evidence of equivalent human-level learning without it. We also have an entire science paradigm that believes it is studying (f)/(g).

Conversely, Figure 7.2 does not include behaviour and adaptation devoid of phenomenal consciousness, that is, lacking the experiential component of cognition. Such non-experiential adaptation reflects innate implicit knowledge in the form of reflexes, instincts or other regulatory systems such as homeostasis. There is one aspect of non-experiential adaptation that cannot be avoided by Figure 7.2. Human cognition is intrinsically able to extract regularity from persistent exposure to an environment via experience (P-consciousness). This is the presupposed property of the Figure 7.2 faculties inside the boxes marked 'novice'. This is the intended basis of learning of the Figure 7.2 kind: regularity extraction and storage mediated by neurological processes that include the generation of experience. The diagram itself, (f)...(i), is not intended to represent physical reality in brain tissue. In reality the components are variously inside each other and different parts may be uniquely localised or overlapping in the brain. The diagram is meant to capture overall function. The process of P-consciousness generation and regularity extraction are innate properties of brain tissue that operate silently in the background, and are not learned. This basic, innate capacity, fundamental to nervous systems of certain kinds, is the physical substrate underlying all Figure 7.2 learning mechanisms[b].

Absent in Figure 7.2 is its application to the learning behaviour of 'organisations' or collections of cognitive agents. Organisational-learning has a long history of examination of the extent to which the cognition of a whole organisation is different to the collective cognition of the component humans e.g. [Wang and Ahmed, 2003]. A way of resolving this issue is to see if a collection of Figure 7.2 cognitive agents (humans) operating as a coherent whole is an instance of a Figure 7.2 cognitive agent/novice. If the physics of P-consciousness were complete, then the answer to this question would be scientifically established. In the absence of definitive science, the organisation as a whole, if it did have experiential components of its own, independent of the humans in it, should be able to exhibit behaviours and learning capabilities of the

---

[b] A future version of Figure 7.2 may want to address this more explicitly.

kind in Figure 7.2 that would otherwise be impossible. It is expected, however, that the only thing an organisation does is operate more efficiently (potentially) for certain kinds of learning. This is what happens in science. Groups of scientist have no collective subjective experiential 'whole' that is more than the sum of its parts. However, the whole can indeed behave more effectively (for certain problems) than the sum of the parts. Future work should provide some clarity on the matter. I need not cover it for my purposes here.

Figure 7.2 could benefit from more attention to learning by non-human cognitive agents. Figure 7.2 may be able to shed some light on the idea of 'self' as a learned construct. The Figure 7.2 cognitive agent definition does not explicitly address situational emotions such as guilt. Figure 7.2 facilitates the scientific study of guilt, but does not explicitly show how such emotions involve themselves in learning. Scientists procedurally rule out these emotions in the creation of new knowledge. Contrast this with a book author heavily invested in these emotions in order to imagistically access the invented world of a novel. Figure 7.2 omits the non-stationary learning capacities resulting from developmental processes, and learning by artificial cognitive agents. Is the internet an example of a Figure 7.2 cognitive agent/novice? Is a single desktop computer running an AI program an example of such an agent? Also missing is an analysis of how the rigour of Figure 7.2 might be applied to scientific testing for certain levels of intelligence or the presence/absence of the faculties of a claimed cognitive agency. Is that computer conscious? Can that animal do science? Can a desktop computer impart explicit and implicit knowledge? How?

Future interesting cases in science include fraudulent science, where a scientist pretends to find an original law of nature. Fraud occurs when interaction with the natural world, Figure 7.2(e), is imagistically, via Figure 7.2(g), manipulated to create a false externalised model Figure 7.2(p) that will subsequently be refuted experimentally. Accidental discoveries can also be analyzed via Figure 7.2. Fraudulent science that accidentally supports a real original, valid science outcome is also covered by Figure 7.2. More generally, Figure 7.2 tells us that it is impossible, except by accident, to pretend to do an act of original scientific behaviour resulting in (or experimentally verifying) a genuinely novel law of nature. If a fraud accidentally discovers something, then it will be validated by original science, and a misattribution of originator occurs, not erroneous knowledge. Original science seems to be intrinsically authentic through inheritance of the authenticity of the natural phenomenon it seeks to describe. If a discovery is already made, an actor can play the part in a recreation of the discovery event. But the acting is not an original act of scientific behaviour. This means that a scientific test for genuine scientific behaviour can be classed as proof of the faculties of the cognitive agent (the presence of Figure 7.2(f), (g), (h), (i) in the tested cognitive agent, i.e. a test for P-consciousness). These issues of the complex neuro-dynamics of learning, testing for intelligence and P-consciousness, fraudulent science and so forth are covered in upcoming chapters.

7.7 Science, scientific behaviour, scientific learning & self-governance

Figure 7.2 can now be applied to the fundamentals of scientific behaviour and its relationship with implicit/explicit models (h/i). The self-governance of tennis was discussed above. As an object in the natural world, the rules of tennis exist, as Figure 7.2(p), in written form. Having learned to play tennis, possibly including exposure to written rules, inside each tennis player is an implicit model (h) and an explicit model (i) learned through the mechanisms discussed above. Tennis behaviour can proceed without written laws. There is no causal relationship between the book of rules and behaviour. The rules operate implicitly as a set of learned constraints within the players, and if you observe behaviour outside the set of constraints, then you have evidence supporting a claim that you are not in the presence of tennis. A Martian scientist observer who knows nothing of tennis may be able to scientifically establish the rule book for tennis, in which case the rules of tennis would become part of the Martian 'scientific laws of the natural world of human tennis'. But then, if an event occurs that seems ambiguously represented or limiting in the rules, then unlike some rules elsewhere in the natural world, the rules can be modified. If new conditions arise, for example the arrival of tennis playing robots that want to win Wimbledon, then review of the rules might accommodate their different physical capacities. This is a rather clumsy way of understanding the role of written rules in self governance. The written rules serve as a benchmark for determining whether a goal (a game of tennis) has been achieved.

Now consider the same idea applied to science. Scientific behaviour is exhibited in a massive diversity of ways by single scientists or huge collaborations of scientists. Yet science is the activity being carried out. If you observe a plumber or an accountant or a tennis player in their normal operational mode, we know it is not scientific behaviour. What does Figure 7.2 say about how it can be that a massive diversity of actual behaviours can be carried out, yet we all recognise it as being scientific behaviour? Scientists go through substantial training, usually in the form of a PhD. Based on Figure 7.2, this means that inside each of us there is learned implicit model (h) and a learned explicit model (i) that enables a human to exhibit scientific behaviour. Of all the vastly different skill sets that are enabled in any individual (h) and (i), what is invariant across all scientists? What is the overall invariant behaviour of single or groups of scientists? For that invariance is the essence of what it means to be engaged in scientific behaviour. That is what distinguishes scientific behaviour from other behaviours.

When asked by a novice Q1 = *"What is the definition of scientific behaviour?"* we scientists will have an answer. As acknowledged, authoritative masters (m), it will be expressed in abstract symbolic terms, driven by our personal explicit model (i) for what we think scientific behaviour is. It will have the form *"Scientists ...."*. The simpler generic answers to Q1 are K1 = *"Scientists discover laws of nature and write them down"* or K2 = *"Scientists explain the natural world"* or K3 = *"Scientists describe the natural world"*. As previously discussed, one of the answers you can find in early literature is K4 = *"Science organises appearances"*[Lewes, 1879; Mach, 1897]. What you will not get is a consistent response of the form K5 = *"I perform descriptive science type 4 as per the Timbuktu accord of 1925."* If you empirically, objectively measure scientific behaviour and average it, the result is very similar to K4. This is covered in Chapter 8.

The answers to Q1 differ markedly from each other and from the examples. Try it. Common forms involve references to objectivity or empirical method or scientific method or falsification and so forth. In contrast, all tennis players will cite the same rules of tennis because they are trained and there is a document to refer to. By analogy, if all scientists had the same answer, then you may conclude that novice scientists have been exposed to the same 'rules of scientific behaviour'. But the fact is the answers you get will vary. Yet scientific behaviour is consistently enacted by all. How can this be?

According to Figure 7.2, each answer to Q1 communicatively externalises an explicit model (i) within the answerer (the master, m). Variability in the answer to Q1 comes from variability in the Figure 7.2(i) model in each scientist. With such demonstrable variability, the invariant core of the scientific behaviour cannot actually result from Figure 7.2(i). If the essentials underlying scientific behaviour were available in externalised form, such as the rules of tennis, then through instructive learning (Figure 7.2(b)) or through rote study (Figure 7.2(d)), a consistent answer to Q1 would result. Science has a very consistent core to its behaviour, so the Figure 7.2(i) explicit model cannot have originated it. Therefore the invariant behaviour must actually be driven by the *implicit model*, Figure 7.2(h). This implicit model (h) came from a Figure 7.2(m) master purely by imitative learning Figure 7.2(c). This is the only explanation, and it can easily be supported by empirical measurement by each of us.

7.8 Summary

This chapter extended cultural learning theory to provide a framework capable of expressing scientific behaviour. It is neuroscience-centric.

The conclusion is that scientific behaviour actually results from an undocumented implicit model learned by imitation of a master (typically a PhD supervisor) during training, who exhibited the behaviour, and demanded it of the novice on pain of the novice's failure to become a scientist. Meanwhile, the novice learned an answer to the question *"What is the definition of scientific behaviour?*" that is an incidental by-product of other aspects of the novice's history.

This mechanism explains how people can become a scientist and operate effectively without having been trained in a formal (explicit) model for scientific behaviour. If there was an explicit model for scientific behaviour (as currently practiced), what would it look like? It has the general form already revealed in Chapter 5. We can now take a look at a more accurate form.

Chapter 8

The 'Law of Scientific Behaviour'

Finally, at a still higher level, there is another set of commitments without which no man is a scientist. The scientist must, for example, be concerned to understand the world and to extend the precision and scope with which it has been ordered. That commitment must, in turn, lead him to scrutinise, either for himself or through colleagues, some aspect of nature in great empirical detail. And, if that scrutiny displays pockets of apparent disorder, then these must challenge him to a new refinement of his observational techniques or to a further articulation of his theories. Undoubtedly there are still other rules like these, ones which have held for scientists at all times.

[Kuhn and Hacking, 2012, Page 42]

Now that the necessary background knowledge of P-consciousness, tacit knowledge and cultural learning is established, a more accurate form of the previously introduced generic statements (= 'laws of nature'), t_n, and 'law of scientific behaviour', t_A, can be measured and assembled. Notionally this can be presented as the result of an explicit empirical study. I say notional because I have not formally carried out the study as a single recent event. Like Thomas Kuhn, what I do is report the measurements of a lifetime's direct observations. Someone may wish to repeat the process in a more formal way. I doubt the result will change much. Individual scientist readers might apply their life experience as an observer of this type, and see how the results compare with mine. I submit my results as sufficient for purposes here.

To do the study, one notionally observes a sufficiently large cohort of scientists and then records what they do. The average behaviour of all the scientists is the 'law of scientific behaviour'. We scientists are not used to being the studied phenomenon, and it may be somewhat disconcerting to realise that in such a survey of behaviour the scientist cannot be asked what they do. That is not what is being measured, which is what we actually do. This prohibition is what is applied everywhere else in the sciences. Our behaviour must be dispassionately and objectively observed. Only then will we be finding what is invariant across all scientists.

What exactly are the kinds of variability to be expected in a record of scientific behaviour? One of these variations is physical location. Should a scientist's behaviour depend on where the science is carried out? Of course not. The *output* of the scientific behaviour may, of course, depend on location. The specific methods and techniques used might depend on the location. But the underlying behaviour is invariant. There are no whales on the moon, for example, so whale science, done with the usual behaviour on the moon, is still science, just not terribly informative.

There are myriad variables like this such as institution type, scientist gender, ethnicity, academic qualifications, career stage, scientific discipline. In each of these categories one must ask not what the actual numbers are, but whether scientific behaviour is critically dependent on the characteristic of the category. If a characteristic is 100% correlated with the behaviour, it is still not proved to be necessary for the behaviour. There is one characteristic that is like this for scientists. It is critical argument with peers. It is not absolutely necessary for science to occur. A lone scientist can still do science. Alternatively, one might consider that a lone scientist is having a critical argument with themselves. The ubiquity of critical argument in science suggests it be left in a formal specification of scientific behaviour, and I have done that here.

Further variability exists in the form of the deliverables. No particular form is a sign of scientific behaviour or coherent with scientific acts. The only fixed invariant is *explicitness* of the deliverable statement(s). Examples of the variability in the kind of media are: book, journal, rock carving. The form of the deliverable statement is also variable. It could be prose (any language), poetry (any language), mathematics (any kind), web site, smoke signals or a blast from a broadcast antenna. The scientific act must result in a deliverable explicit scientific statement, but the transmission medium of the scientific statement is irrelevant in that it does not determine the existence of the statement or the accuracy of the statement. It merely determines the level of convenience of access, storage, portability and comprehension. The invariant fact across all science is *merely the fact of the explicit statement*. There shall be an explicit statement. If you don't make a statement then you've not done scientific behaviour. Consider this fact within the context of what I am doing here, now. I am trying to formulate a scientific statement about the invariant attributes of scientific statement production. If I do not create a statement, then I cannot claim to have contributed to a science of scientific behaviour.

The scientific outcome is invariant to the usage of particular physical devices, skills, processes or procedures. A scientist might sit and observe birds or use a particle collider, a PCR machine, surgery or a telescope. The outward signs of a scientific act cannot, in themselves, prescribe that any of these things must have been involved. Scientific statements arise, and while their quality and scope may be determined by a particular technique, no particular devices, skills or processes are essential in defining the act of producing a scientific statement.

The same argument applies to the use of particular mental skills or processes or procedures. There is a bewildering array of such things. If you sample the literature on science you will find, amongst many others, scientific behaviour includes the use of: statistics, decision/rational choice theory, theory formation (theory-theory), analogical and imagistic reasoning, mental modelling, visualisation, spurious correlation handling, anomaly resolution, computational methods, in-vivo/in-vitro, problem solving techniques, serial/parallel processing, heuristic searching, genetic searching, machine learning, induction/abduction/deduction, mental rut handling, confirmation bias, coordination of evidence with multiple theories, conceptual change, the 'Aha'/insight moment, idea incubation, creativity vs. reasoning, handling and interpretation of insufficient or disconfirming or mixed or ambiguous evidence, metaphor, limiting case experiments, thought experiments and so forth.

Once again, no particular member(s) of this list is prescribed, nor are there others that might be explicitly proscribed. Scientific behaviour, in the sense to be formally captured here, is invariant to all these things. The usage of these things may determine the accuracy, quality, scope, reproducibility and defensibility of a scientific statement, but ultimately the scientific act does not causally necessitate any particular kind.

Scientific behaviour only requires a normal, awake, alert healthy human brain directed at behaving scientifically. It requires a brain that is plastic and able to create/modify statements based on experience i.e. *P-consciousness* as scientific observation (of the kind depicted in Chapter 7). Degrade or eliminate P-consciousness and the science outcome will degrade or disappear. Note that this position does not deny that scientific behaviour may be possible without P-consciousness. The position merely recognises that, to date, there is no measurable instance of scientific outcomes produced without it. Therefore P-consciousness, *prima facie*, is necessitated (causally antecedent) for scientific behaviour.

Scientific behaviour (statement production) is invariant to motive in the sense that it does not matter what particular *kind* of motivation resulted in a scientific statement. The fact of the scientific statement is the only requirement and implicitly evidences a motive of some kind. The invariant component of this is that there shall be sufficient motivation. Without it there will be no scientific statement. Scientific behaviour is therefore invariant to all the previously listed human foibles that involve themselves in motivation: personality type, social/political circumstances, competition, life goals, authorities, prestige, entrepreneurship, tradition, fashions, eminence, mentor, preferences, prejudices, favourites, personal styles such as 'back scratching', collaboration, group/team science, use of secrecy, branding, peer association, relationships with sources of funding.

Scientific behaviour is also invariant to developmental issues such as the child-scientist, the gifted, and the mid-twenties discovery cusp. The subject area, effectiveness, timeliness and quality of the result may be a function of these things. However, the basic behaviour – the production of scientific statements – does not necessitate any particular stage or sophistication of human development. It does, as noted above, require that the scientist is a healthy, awake, alert human, at least so far. No non-human entity (such as a chimp or a dog or a robot) has made scientific statements of the kind targeted here. As already noted, intelligence revealed by toolmaking does not exhibit scientific behaviour, but merely a level of intelligence and behaviour able to solve problems. The explicit, portable scientific statement is what makes humans (so far) the only scientists. There are two special cases considered later in more detail.

These are (a) the intelligent alien scientist and (b) the robot scientist (already claimed operational in various places around the world). These two special cases will temporarily be set aside because they will not change the outcome.

Scientific behaviour is also invariant to the accuracy of the scientific statement. This sounds strange, but it is a brute reality that most scientific statements, in the sense of deep history, have been overturned or superseded for a variety of reasons. What was once a scientific statement considered a bastion of truth for a century (for example the phlogiston theory of combustion) can later be seen to be so wrong it looks bizarre. Yet at the time the particular scientific statement was as scientifically acquired and posed and as predictive as it could be. Therefore, scientific statements are, in fact, invariant to their sense of being 'correct' or 'right' or 'truth' or 'fact'. I am about to formulate a new scientific statement about scientific behaviour (production of scientific statements). I must, therefore, insist that it will also lack the same absolute sense of being 'correct' or 'right' or 'truth' or 'fact'.

All these aspects of scientific behaviour capture the signal and the noise in our measurement of scientific behaviour. What is that essence of scientific behaviour that remains once all the noise is averaged out? It is the 'law of scientific behaviour' called t_A in a previous chapter.

8.1 Scientific behaviour and the scientific paradigm

If we are to account for the big picture of the evolution of scientific behaviour (production of scientific statements) then the long term paradigm-to-paradigm dynamics could be thought to have caused an unwritten 'law of scientific behaviour', t_A, also to change over time. If so, then t_A is non-stationary even though it may not have been written down. One might trace the evolution of t_A and thereby gain insight into how humans started with no science, began to seek regularity in the world and document it, thereby becoming predictive with a primitive (proto-) t_A. Ultimately, our modern t_A accesses natural regularity so well that it has resulted in the technology of today. Yet it remains undocumented.

Consider the only-just-linguistic human of, say, 100,000 years ago who may have been able to write down a proto-t_A such as "*Everything the tribal elders say is a regularity in the natural world.*" Such a 'law of scientific behaviour' would stand or fall on the predictive capability of the tribal elders. During the era of the ancient Greeks, a t_A such as "*knowledge is captured natural regularity supported/acquired through experience of the natural world*" may have been the next evolutionary step in t_A. There was no need to write it down, so the knowledge is implicit in the minds of the artisans and empirics of the day. This is not specific or nuanced enough to capture scientific behaviour for the purposes of the third millennium, where an artificial scientist is a real possibility. Modern scientific behaviour stabilised in its current implicit form sometime in the last 350 years or so, perhaps longer. Someone else can sort out the exact dates (if there can be such a thing, and Kuhn advises this kind of date-stamping is fraught). As has been discussed at length, it is an implicit 'law of nature' but it has, in spite of being handed down by imitation, been remarkably stable for many lifetimes - well out of living memory.

Having recognised the long term variability in t_A, the capturing of its contemporary form is not a work of history, it is a work of science in the present day and looks for a statement of regularity in nature, as science does when enacted anywhere else. By constraining our examination for scientific evidence of a t_A to the modern era, the sought-after t_A can be claimed invariant to scientific era/paradigm in the manner that all other laws of nature acquire precedence over their predecessors. We can do this in the full knowledge that it must have changed over time in the ways exemplified above.

8.2 Down to business: t_A

In an ideal science world, to proceed from this point, what we need to do is simply sample scientific behaviour over the modern era. With a sufficiently large sample set, all of the variability can be averaged. What remains is the invariant, unchanging core of scientific behaviour. In the real world of this book, what I am going to do is, based on all the previously listed constraints, simply write down what my experience of scientific behaviour suggests is t_A. I have used my own 'sample' of all the scientists and engineers I have ever known and read about. The scientist/engineer reader is encouraged to do the same. Plug your own data into the upcoming t_A and see how it fits.

Before we do t_A there is the matter of the generic form of a 'law of nature', which, you may remember, has not been finalised yet. We left it in this form:

$$t_n = \quad \textit{Some statement about the natural world.} \qquad (8.1)$$

A more useful form of t_n can also be assembled from each scientist's experience of being a scientist. I have done this, and have found that every scientific outcome over the modern era could, if required, be standardised by recasting it into the following form:

$$t_n = \quad \text{The natural world in} < \textit{insert context} > \qquad (8.2)$$
$$\text{behaves as follows:} < \textit{insert behaviour} >$$

This is a straightforward enough development. It enables us to specify what the natural regularity is and the context in which it applies. Try it for yourself. Take a well known or obscure natural regularity and cast it in the form of equation (8.2). Quantum mechanics could be recast to fit in the form of equation (8.2). You can even create a law of nature about the items on your desk. Such ad-hoc, trivial regularities are not particularly useful or lasting or publishable. Nevertheless, the natural world of trivial states of affairs is just as amenable to capture by equation (8.2) and are as predictive as any other, just not the sort of thing you find in the scientific literature. Scientists have bigger fish to catch. It is explicit (you write it down) and it can be passed on to others (portable). That is the simplest, most practical way to view t_n.

The laws of nature that we produce have already been argued to be purely descriptive, making them predictive of how the natural world appears when we look (scientifically observe). That is, the natural world will appear, when we look, so that it is never inconsistent with t_n. This is the most intellectually un-laden, un-presupposing form of a law of nature that it is possible to construct. No matter how complex or mathematical, we don't claim it to be 'right' or 'truth' or 'fact' or any of the other obfuscating labels found attributed to such statements in the literature. We don't claim a particular t_n to be unique. It only needs to be intelligible to an appropriately trained observer that can then hold the regularity to account when it confronts the natural world. A t_n makes no other claim about the relation between the regularity and the natural world. This position does not mean that there is no other formal position that may claim or prove a relation between the natural world and such t_n. It merely means that from the point of view of the scientist constructing them, such claims (i) change nothing and (ii) whatever the extra claim is, there is no evidence in t_n that supports it. This is the most theory-unladen position that we can empirically support.

We can do better than equation (8.2). Perhaps the observer dependence of equation (8.2) is not explicit enough. This loads the words 'behaves as follows' with an implicit assumption of the need for an observer. The natural world, in the context of a (scientific) describer of its regularities, can only 'behave' in the eyes of a (scientific) observer. That being our objection to equation (8.2), we might consider reworking it as follows:

$$t_n = \quad \text{The natural world in} < insert\ context> \hspace{2cm} (8.3)$$
$$\text{will be observed to be consistent with:}$$
$$<insert\ behaviour>$$

In this way, the role of the observer is made more explicit. Some scientists may prefer this. The fact is, it is not my position to declare a full or final t_n. That is the role for as yet undefined bodies of scientists allocated to the task on behalf of science as a whole. That is what a *governing* body of scientists can do for science; exactly what the governing body for tennis does for tennis. Once we have one, then we can argue about merits and all the surrounding issues. Remember, this is not philosophy: equation (8.2) is an empirically measured/verifiable 'law of nature' that captures what scientists produce. I challenge all scientists to make the same measurement and critique equation (8.2). When you do that, you will be amongst the first scientists ever to actually make that measurement and do that critique.

We now have a working t_n, which tells the 'what' of scientist's output. We can now look at t_A, a particular t_n which gives us the 'how' of their production. Based on my own experience of scientists and engineers, when I measure what we do and distil out the signal from the noise, my attempt at stating the invariant core, expressed using the t_n form (equation (8.2)) as a template, is

$t_A =$ The natural world in < *the context of a human* (8.4)
being scientific about the natural world >
behaves as follows: < *to create and manage the*
members of a set of statements of type t_n, each of
which is a statement predictive of a natural
regularity in a specific context in the natural
world external to and independent of the
scientist arrived at through the process of
critical argument and that in principle can be
refuted through the process of **experiencing**
evidence *of the regularity*>

This is not claimed *the* t_A. Just like equation (8.2) or equation (8.3), equation (8.4) is not claimed unique or final or even correct. It simply what I measure and report. Once again, it is not up to me to determine the final form of it. That, too, might be thought a matter for a governing scientific body to make formal based on evidence and critical argument. Just like any other 'law of nature'.

There are many features of equation (8.4) worthy of discussion. Equation (8.4) presupposes a scientist and the scientist's cognitive faculties, in particular the ability to 'scientifically observe' (P-consciousness, the 'experiencing evidence' clause in equation (8.4)). This is exactly what scientists have traditionally presupposed all along. As a result it is easy to predict that t_A will fail to explain a scientific observer or how science is possible in the first place. The circular logic of this is more visible by analogy, such as "*to claim that observations explain an observer is like saying that telephone conversations explain the phone system*". It is a nonsensical claim.

Statements of kind t_A are all merely descriptions, and contain no explanation as to why it is that way. Therefore the failure to explain the observer is merely a special case of a general explanatory failure intrinsic to statements produced by the t_A procedure.

Equation (8.4) is not 'scientific method'. Scientific method emerges as a more formal optimisation of the process of equation (8.4). This results in the more accurate, timely access to a t_n. Furthermore, observe that the working life of any individual scientist may not appear to be the process t_A. In reality, portions of t_A may actually carried out by members of a team of scientists. In the end, every part of t_A must be satisfied or scientific behaviour has not happened. However, a single scientist could exhibit behaviour t_A. Equation (8.4) does not prescribe the life of a scientist. It prescribes the necessary activities constituting scientific behaviour that results in a scientific statement. The statement t_A results from the overall statistically measurable ensemble average behaviour of scientists as they go about accessing regularity in the natural world, abstracting it into a statement and then depositing it onto (or removing it from) the great pile, T, of all such statements. That pile of statements is the entire scientific publishing system, regardless of the kind, form and accuracy of the statements therein. Statement t_A reflects an objective view of scientists as a form of cognitive agency directed towards specialised ends (new predictive statements) that are unique in that they are objectively testable and independent (if the behaviour is successful) of the agency (the scientist) that produced them.

Note the intrinsically self-similar nature of the t_A/t_n pair. t_A is a statement of type t_n about the process of construction (scientific behaviour) of statements of type t_n. This is the natural and expected property of a regularity statement that is a process of regularity-statement construction. It is claimed here that such a characteristic is a necessary attribute of a correct system of scientific statement production. This self consistency and self-similarity brings with it the property that it accounts for our ability to evolve from a state of having no science, to having science, yet have no formal, explicit definition of what it is to behave scientifically. It allows for a t_n to be written down without an explicitly written down/trained t_A. This permits a long era of experimentation with various implicitly held t_A that are passed on by imitation from mentor to novice. Of course, all of it is ultimately grounded in the innate biology of cognition in humans.

As already discussed above, note that the equation (8.4) requirement for critical argument (peer review) is not a necessary component of t_A. Formal science can be done by an individual scientist in isolation. The quality and effectiveness of the process may be affected, but the basic procedure is sound. The critical argument clause was retained because it had statistical significance that made it a characteristic worth keeping inside t_A. It is what you get when you measure the behaviour. Perhaps a future version of it may not include it. Critical argument can be thought merely a correlate, not a critical dependency (a necessary prior). Future versions of t_A may change that status.

As a statement of kind t_n, t_A inherits all the value-free characteristics listed above. Apart from the application to the natural world of scientific behaviour, there is nothing special about it. To put the t_A process of

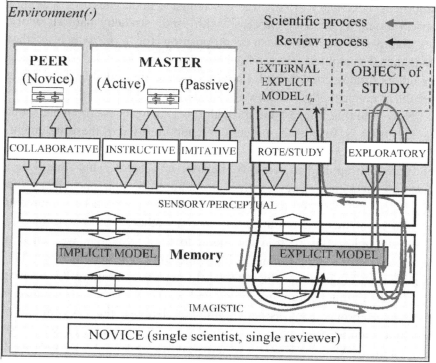

Figure 8.1.   A single scientist produces a paper, one or more reviewers do critique and it is returned for modification. When revised and stable it enters the journal system. For details of the unlabelled boxes and their meanings see the chapter on cultural learning.

equation (8.4) in the cultural learning terms discussed previously, including the peer review process, consider Figure 8.1. It is based on Figure 7.2 and contains the process of becoming a scientist and then enacting routine science as per equation (8.4). A scientist goes through all three types of learning (collaborative, instructive and imitative) as initiation, during which the novice scientist prototypes their ability to enact t_A, which is shown used in practice in the rote/study and exploratory components. In the absence of a formally expressed, trained t_A, the substance of t_A is captured by the implicit/tacit knowledge part of the diagram.

In this way, behaving as per t_A results in a new t_n. The Figure 8.1 arrows depict the overall flow of activities and information that a scientist enacts when a new statement is constructed. In classical empirical science, first the scientist undergoes a looping observational/interactive process with the studied natural world (exploratory). This results in a proposition for a t_n, which is then constructed in explicit form and delivered as 'rote' knowledge for review by others. It may come back for alteration or be discarded or accepted as per the dynamical physical flow depicted in Figure 5.1.

Now note that I have gone through this exact process to create t_n and t_A. This book is, for me, the process of externalisation of these statements for review and critique by others. The fact that we've never had this kind of formalism applied to scientific behaviour is moot. This should be a natural expectation – we scientists are natural, regular phenomena like any other, and we need to accept that our behaviour can be captured in scientific statements and trained, just like any other.

To complete this section and point to future chapters we can now revisit and upgrade the Figure 5.1 dynamics of scientific behaviour. This is shown as Figure 8.2, which explicitly depicts the Figure 8.1 looping (exploratory) interaction with the natural world. What has been added is the special terminology (as per Figure 4.2) denoting the environment in which a scientist encounters the natural world *from within,* as part of the environment. Also recognised explicitly are the collections T (successful, accepted statements) and H (unsuccessful, ejected, hypothesised or untested statements), duplicated here for convenience:

$$T = \quad \{t_A, t_1, t_2, \dots, t_n, \dots t_{N-1}, t_N\} \quad (8.5)$$

$$H = \quad \{h_1, h_2, \dots, h_m, \dots h_{M-1}, h_M\} \quad (8.6)$$

In equations (8.1)/(8.2)/(8.3), (8.4), (8.5) and (8.6) we have a set of prototype statements from which a more widely accepted form may emerge. All of this is backed up by cultural learning theory, and is, in the end an *empirical outcome*. It was literally *measured* using itself as a process. Reasons for it being important enough to earn this investment of effort are the subject of this book from the next chapter on.

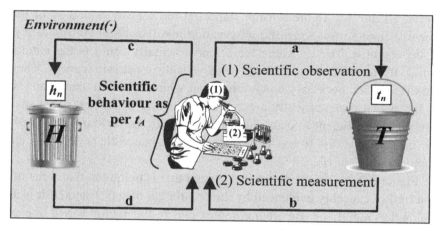

Figure 8.2.   The dynamics of scientific statement production.

There are important subtleties in Figure 8.2. First, notice that the physical embodiments of sets T and H are embedded in/part of the environment just as much as the scientist. This has an important implication for the process of science formalised above: set T and set H are the products of a complex sequence of natural causality, just like any other phenomenon in the natural world. This is meant in the strictest sense of causality. Consider the statement $t_{1004} = $ *'The number of aardvarks in region X is 1234±50'*. The scientific evidence for this statement is literally a cohort of aardvarks observed by a scientist. Now ask, in the natural world of the behaving scientist, what is the scientific evidence of scientists? The answer is the sets T and H. I mean this literally. These are the equivalent to the aardvarks in the example. Furthermore, because of the nature of the t_A process, the statements in set T are literally constructed to be independent of any particular scientist. The natural world depicted by a t_n operates consistently with t_n without the scientist being present. Obviously the only t_n that demands the scientist be present is t_A! Even then, the process is independent of any particular scientist.

The next important nuance in Figure 8.2 is the distinction between scientific observation and scientific measurement. In scientific measurement, we scientists may use our basic senses or may augment them with instruments. This involves the natural causality in the external world, all the way to the sensory impact at the sensory boundary of the scientist. In contrast, scientific observation involves the P-consciousness of the scientist. This is where the scientist actually contacts the natural world. If you take either of these away, scientific behaviour stops. They are the verified necessary conditions upon which scientific behaviour is, as far as can be empirically ascertained, critically dependent. They are, however, insufficient to cause an act of scientific behaviour. That needs the hosting scientist to exist, to act in accordance with t_A, and to then populate set T (or H).

Figure 8.2 is a single dynamical system that is unique in that it results in a trail of causality evidenced by the population of sets T and H. It is as real and as natural as any other complex part of the natural world like an ecology or a galaxy. In this way, we have completely naturalised the process of science and made it indistinguishable, in principle, from any natural process. Its only uniqueness is that it is a specialised natural phenomenon directed at producing a special kind of statement about the natural world.

8.3 Causality, apparent causality and t_n

Before we look at causality within scientific behaviour we need to clarify it generally. The concomitance and sequentiality of phenomena are built into the statement t_n clause *<insert behaviour>*. These statements are literally and only constructed to be predictive of appearances. That is all that is claimed. The predictions themselves are our only point of contact with causality. Call it 'apparent causality' because the statements only expose how causality is evident in the appearances of phenomena. For example, if t_n was Newton's 2nd law, $f=ma$, then the logical consequences of it, informing us of how the world will appear, are built into the statement. The statement t_{1004} about aardvarks (above) tells us to expect the logical consequences of there being aardvarks in region X. Things such as aardvark food, predators, nests and so forth. Those t_n about heat transfer thermodynamics and water phase transitions may connect us with a prediction of car windscreen condensation. That is the extent to which we can claim to understand windscreen condensation in the sense that is traditionally held as 'explanation'.

In this way, by connecting different t_n statements to each other we can track events and predict events descriptively. None of this process says anything about why the events unfold the way they do. Nor do we have the right to impose any such responsibility on the statements. We need discuss it no further because such discussion goes beyond any aspect of the natural world for which statements of kind t_n can be claimed to have something to say. It is a logical endpoint. The statements contain no causal mechanism and we are not entitled to impute any into them. Notice that we have remained entirely within the nomenclature and concepts of the t_A/t_n system, which is an empirically determined framework for construction of statements stored as a collection called set T. This is all we need to know to use the framework to handle 'apparent causality'. It is as simple as it gets.

The t_A/t_n system is thereby admitted as irrevocably mute and impoverished in relation to causality (that which necessitates what we observe, and also how we can observe in the first place). In anticipation of coming chapters, please ensure that this impoverishment is not treated as a proof that causality (yes, true causality) is unassailable by science. That would be to make the grand mistake that has blighted science and caused hundreds of years of arguments. That mistake is to assume that the t_A/t_n framework is the only way scientists can construct statements about the natural world. Later we will show this to be an erroneous assumption.

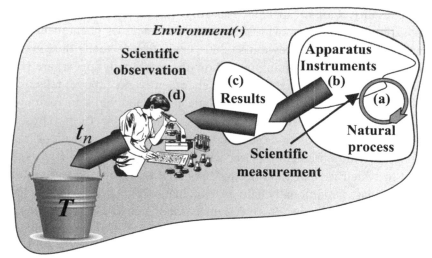

Figure 8.3.   The chain of apparent causality of scientific statement production.

Now let's tease out the 'apparent causality' within scientific behaviour itself. This is shown in Figure 8.3, where we see the studied natural world (a) interact with some kind of instrumentation, (b), which then presents results, (c), to a scientist operating as per t_A. There, the measurements are scientifically observed, (d) (pass through the P-consciousness of the scientist), are processed mentally/creatively to result in an explicit statement t_n that ends up in set T (usually the literature somewhere). Every part of this chain of events is a necessary component for science to result. Take any one part away and the entire chain stops, and set T (and H) contents remain unchanged.

Consider a circumstance where a 200 year long experiment involves measurements (Figure 8.3(a)), in remote deep space that are made and travel back to earth, where the originating scientist is no longer alive. No set T impact occurs until scientific results (Figure 8.3(c)) pass through the P-consciousness of a scientist (Figure 8.3(d)). Each step is necessary, and it doesn't matter whether it is done by a team or by an individual. Each and every step in the chain has to occur or set T/H remains unchanged.

As discussed above, in nature a claimed chain of (apparent) causality is established through arguments based on statements of type t_n used to reveal the necessity of each link in the chain of (apparent) causality. For example, the neutrino was claimed discovered because a cogent argument was made that the causal descendants of a neutrino, and only a neutrino, were measured and that measurement was then observed. No-one has ever directly observed a neutrino. A statement about the natural phenomenon we call the neutrino arose through exactly the process of Figure 8.3. In the chain of apparent causality argued as evidence, every link in the chain has to be present or the neutrino would not be scientifically observed as part of the Figure 8.3 process. Therefore, the instrumentation provides a succession of necessary antecedent events that began with the existence of a neutrino. The observation is then claimed evidence for every single link in the chain, including the neutrino. That is, causal descendants of the neutrino evidenced the neutrino.

That being the case, and back in the chain of causality of Figure 8.3, what does the contents of set T evidence? It also is the result of a tortuous chain of apparent causality of exactly the same kind as that of the neutrino. Therefore, by analogy, set T delivers scientific evidence of all the myriad causal links involved, including Figure 8.3(a) ... (d). Set T, it can now be said, evidences each of Figure 8.3(a)...(d), *including the P-consciousness of the scientist*, the singular and only place where the human scientist literally encounters the original natural process Figure 8.3(a).

Think about that. We now have some kind of argument for scientific evidence of P-consciousness. It is witnessed in the population of set T by the process described by t_A, which is specifically designed to make statements *independent of the P-consciousness of any particular scientist*. Set T is therefore objective evidence in its own right that, amongst other things, includes evidence of the subjectivity of the scientist in an objectively verifiable way. It makes no difference that we don't understand how P-consciousness arises. That it is something upon which scientists are critically dependent is child's play to verify: take away or degrade P-consciousness and watch set T population stop. It makes no sense to demand, via t_A, that a scientific observer experience evidence of a regularity using P-consciousness, and then to deny that P-consciousness has been evidenced. This is the kind of anomalous, systemic denial that has kept P-consciousness unassailed by science for hundreds of years. The t_A/t_n framework allows us to reveal that anomaly. In this way, the claim that *'there's no scientific evidence for consciousness'*, which has become the classic retort used to shun consciousness for so long, is paradoxically revealed as being within the behaviour of ourselves as scientists. This will be discussed further in Chapter 12 on scientifically testing for P-consciousness.

It seems a little clearer now how the issues surrounding consciousness in science have proven difficult for so long. It is also ironic to note that the road to clarity came about by scientifically tackling scientific behaviour as a natural phenomenon like anything else, something implicitly avoided by dint of cultural norms, not because it was impossible.

8.4 Statement dynamics

I am not suggesting that the entire output of science must be recompiled into $t_n/t_A/T/H$ framework form. It is posited as a way of encapsulating what we do, for examining history, and for seeing the future. In the sense of history, the $t_n/t_A/T/H$ framework facilitates a more formal systematic view of scientific paradigms and the shifting thereof. Consider a few cases of set T dynamics:

(1)   A t_n can be removed from set T (example – phrenology, astrology).

(2)   A t_n <*insert behaviour*> clause can be replaced (e.g. Einsteinian kinematics replaces Newtonian).

(3)   Through a bifurcation of the equation (8.3) t_n <*insert context*> clause, a specialised form of a more general t_n can be created that is better at handling a novel/particular context.

(4)   An entirely new t_n can be deposited into T (e.g. t_A, or the arrival of Kuhn's *Structure*).

(5)   A t_n can be refined (e.g. to create a more accurate physical constant within it).

In principle, one t_n can directly reference another so that the lineage is explicit. Each change can, to some extent, speak to the paradigm. Set T is actually a database, and searches keyed on <*insert context*> and <*insert behaviour*> clauses could be very revealing, especially if they included people, locations, dates and so forth. Eras of relative stability (or consistent increase) would reveal paradigms. Moments of minor disruption (minor revolution) or sudden change (major revolution) would be detectable. It will never happen formally like this, but this kind of 'seismology' of set T is essentially what Kuhn's *Structure* was about, and it all ultimately rests on the seismology of brain neurodynamics. Cases 1, 2 and 4 fit the Kuhnian category of a revolution, and might be regarded as a shift in paradigm (perhaps confined to such a refined scientific sub-discipline that it becomes largely invisible elsewhere). Cases 3 and 5 are consistent with the 'articulation of paradigm' and of 'normal' science elaborated at length by Kuhn.

The claim that science is a purely gradual cumulative activity is implicitly disproven by the framework, which is in line with Kuhn's arguments. The set T has contents, and those contents may be classed in some way as an 'amount' of science quantified in the number of statements or in the diversity or otherwise of <*insert context*> and <*insert behaviour*> clauses. Whatever the metric, its generally increasing value is expected, by virtue of the increasing number of science disciplines that are pouring output into set T, all of which is empirically proved by the ever-increasing complexity and kinds of technology based on it. Each new set T member also asks more than one new question, and reveals more ignorance. On this basis alone, set T must be ever-increasingly occupied with ever more diverse statements. Nevertheless, fundamentally, set T content is finite, non-zero and dynamic in both directions up and down, and the more revolutionary the statement, the more instability and change there is in the contents of set T. If all humans suddenly disappeared from Earth, then set T would suddenly become inert; frozen in time. If the products of science were later discovered in their inert state by, say, an alien scientist, then they could conclude that there were scientists on Earth once.

8.5 Statements and objectivity

The $t_n/t_A/T/H$ framework has no explicit demand for objectivity. The basic demand for objectivity is met through the demand to 'experience evidence'. The other demands associated with objectivity are derivative and serve to improve observer-independence, accuracy, timeliness and so forth. The essence of the framework is that of formally necessary attributes upon which predictive utility is critically dependent. The demand to 'experience evidence' and explicitly report it demands that something be 'objectified out' of subjectivity.

8.6 Statements and 'law', 'theory' and the like

The presence or otherwise of symbols expressed in the formal manner of an abstract mathematics or logic is moot in the $t_n/t_A/T/H$ framework. Their presence is merely an indicator that the natural world's *appearance* has, for reasons unavailable under the scope of the framework, usefully cohered with the dynamics and content of a formal symbol system of some kind. Within the framework, all that has to happen is that an appropriately trained scientist, using a statement t_n from set T (in the literal sense of its inhabiting the scientists brain) becomes predictive in a useful context (which also inhabits in the scientists brain at the time of a t_n's use). Statements of kind t_n are to be regarded as nothing more than captured observational regularity used as a 'lookup index' to the future brain dynamics of a hosting scientist. This neural mechanism is to be detailed in the next chapter.

The 'lawfulness' of a statement or the 'theory-ness' or the 'fact-ness' or the 'truth-ness' of the statement are concepts that have no place in the framework. Note that I have also consistently tried to avoid, unsuccessfully, the phrase 'law of nature'. That phrase is gestalt-poisoned. It suggests that the regularities and formalities in a statement are somehow literally exerting control over the natural world. That kind of allusion needs to stop. The use of the word 'statement' has been established here for that very reason. Its actual usage is aligned with the word 'utterance'. You will note that in Kuhn's *Structure*, these unnecessary, obfuscating, misdirecting words are used throughout, albeit cautiously and with wisdom. I have no need for them here. The entire story could be written without them.

8.7 Fake science

In the cultural learning chapter we came across the actor, a pretend holder of knowledge of X, as an instructor teaching X in an instructive learning context, and contrasted that with a novice acting the identical role as form of imitative learning, both to teach X and to learn X. In both these cases the novice can acquire some imperfect but useful sense of the necessary held knowledge underlying skill in X or skill in teaching X. In both cases (the acting instructor, and the acting novice) there is no fully internalised knowledge of the target X, yet the learning of X can be seen to be part of the process. A novice being asked to teach X can be an effective way for the novice to learn X. At that time the novice is, to some extent, an actor.

How do these ideas extend to scientific behaviour? There are many famous examples of scientific discovery that turned out to be fraud. Scientific behaviour was claimed to have been carried out, and yet in reality it was not. During the process, the scientist(s) involved were, in fact, pretending/acting the behaviour. That being the case, some part of the Figure 8.3 chain of apparent causality was corrupted or did not involve the natural world that was claimed to be under scrutiny. Contrast this with a genuine mistake. Genuine mistakes occur in the mal-operation of the scientist's brain. A behavioural/perceptual aberration of some kind has occurred that resulted in an outcome that would have otherwise been different. It seems, at this level, that the physical signs of fraud and mistake are the same. What is different is the motive: clearly a brain phenomenon. Mistakes can be made at any stage of the Figure 8.3 chain of apparent causality. So can fraud.

Regardless of fraud or mistake, a misleading statement t_n is deposited into set T. What has been compromised is the predictive utility of the t_n. It is no longer about the natural world, and will suffer the fate of non-reproducibility. It cannot reliably be used as a predictor of future states of the natural world, or the results upon which the statement was based will subsequently be found to disagree with measurements, the original measurements will fail to be replicated, or the natural world itself might be found manipulated to suit the desired outcome. These are the points of weakness throughout Figure 8.3 that betray the malformed scientific act.

Now, let's turn all that around and consider the genuine article. One of the requirements of science is that it is fundamentally aiming at the unknown. That is, it is all about encountering novelty in the natural world. Remember that Kuhn advises that history reveals paradigmatic science in which novelty is apparently not required and is, indeed, shunned. According to Kuhn, articulation of paradigm is not about novelty, it is about getting the natural world to fit in with the already-established view of the paradigm. But this is a skewed view of novelty. If you look at t_n, novelty can be either in the natural world behaviour itself, or in the novel context. Even within a paradigm, new cases, new problems, new puzzles pit the received view against nature in novel contexts. Novelty must be thought of as indicated by the differing t_n in set T. Let's take another look at t_n (Equation (8.3) for no particular reason):

$$t_n = \quad \text{The natural world in } < insert\ context> \quad (8.7)$$
$$\text{will be observed to be consistent with:}$$
$$<insert\ behaviour>$$

For a scientist mandated to encounter novelty or cease being a scientist, there are obviously two main kinds. Either (1) *<insert context>* changes while *<insert behaviour>* remains constant, or (2) *<insert behaviour>* changes while *<insert context>* remains constant. Paradigmatic science does more of (1), and (2) arises and involves itself more in matters of paradigmatic shift. In either event, the scientist is encountering and dealing with novelty. If not, then the statement is already in set T and there is no novelty. Sometimes two independent scientists, unaware of each other, deposit the same t_n in set T. But that is more a matter of rights of priority of discoverer, not any indictment of the originality (novelty) encountered and handled by the individual scientist.

It is in its fundamental contact with novelty that fakery in science distinguishes itself from fakery in other areas. Faking a knowledge of X in the teaching contexts discussed above is fundamentally different to fakery/pretending in science because, by definition, there is no master of knowledge X. The novel knowledge does not exist. There is nothing to fake. The scientist is to become the master, and is charged with creation of knowledge X in the first place. Only then can knowledge of X be faked in the sense of pretending mastery of it. Faked science is fundamentally different in that the scientist was supposed to originate/validate knowledge through exposure to the natural world, and instead has originated knowledge via another means. The originality is corrupted. Genuine novelty in the natural world has not actually been encountered.

When put this way, it starts to become clear that evidence of fakery in science and evidence of fakery elsewhere have a fundamental difference. In science, the outcome is arbitrated by the natural world and reproducibility of results (predictive utility), and is additionally mandated to occur in a context of an encounter with genuine novelty (a novel t_n is afoot). That is, faking a genuine act of scientific behaviour (discovery) is fundamentally impossible. By definition, nobody else knows the novel result of a known context or the novel context of the known result. The scientist is there to encounter the novelty and capture the knowledge.

Perhaps the word impossible is a little too strong. Consider the case where a genuine mistake based on erroneously acquired measurements or poor reasoning resulted in the creation of a valid scientific statement. This has probably happened. That being the case, set T has had its membership increased, and it may as well have been an original scientific act that was behind it, for it withstands the scrutiny of others despite it having been accidentally acquired. Scientific behaviour was still enacted in good faith by the accidental scientist(s) involved. Fakery did not occur.

The robust resistance of set T population to fakery, an honesty inherited from the utter impartiality of the natural world itself, means that if a scientific act producing a genuinely novel statement (that no-one else has knowledge of) is observed, then that act is not faked. It may not be particularly useful, but it's original. If it is faked, then set T predictiveness has been compromised and the result will become obvious when set T is used. If the predictive utility of a t_n is compromised, either the context of its applicability or the regularity itself must be altered until its predictive utility is restored. This is the reason why astrology, for example, is no longer in set T.

8.8   Robot science

The robot science discussion has many similarities to the 'black box' argument in Chapter 4. It turns out that there are many 'black-box-scientists' around the world already. These installations are called automated science, sometimes using robots, sometimes simply using computers. Are they robot scientists? What these installations do is take advantage of repetition and the timeless patience of machinery to explore the natural world in a context that swamps a human with vast amounts of natural detail and variations on a theme. Is this science being done by a robot or has the human science routine merely been automated? Based on the popular media presentations of such things, one might get the impression that human scientists may become less necessary as we march into the future. Are these installations carrying out scientific behaviour as we have depicted it?

We can look at this by examining every part of the chain of apparent causality in Figure 8.3. Humans are involved in every part of the chain, all of which resulted from the enactment of human behaviour captured by t_A. We decided on the aspect of the natural world to study. We organised the measurements to be made. We collected the measurements. We observed the natural phenomenon through the lens of the measurement technique. We distilled apparent regularity from our observations. We created some kind of abstraction, one or more statements of kind h_n, capturing the apparent regularity, and we went through the peer review process, which, if found acceptable by the relevant community, is transformed into a t_n and deposited in set T. Based on this alone, clearly the automated installations attracting the name 'robot scientist' cannot be claimed to be doing what humans do.

We can accept that some of the robotic science involves elaborate computer-based searching for patterns, and candidate patterns of considerable and unexpected novelty can be constructed by the robots. When that happens, is it an instance of the causal chain of the previous paragraph? It doesn't take much analysis to arrive at an emphatic no. In fact it is obvious that the only parts of Figure 8.3 covered by robot scientists are (b) and (c). These installations are elaborate instruments. The robot did not design itself(!), did not choose what part of the natural world to study, did not create the searching/data mining and hypothesis generation techniques that are used, does not accept or reanalyse the actual final results for their suitability as scientific outcomes, and plays no part whatever in the peer review process. In particular, at no stage can anyone claim that the robot has done any scientific 'observation'. The measurement process of Figure 8.3(b) cannot be assumed, by any law of nature in our set T, to have been involving P-consciousness (within the robot). Nothing of that nature is explicitly designed into the instruments, and indeed the entire affair could be carried out by scientific instrument makers that have no awareness that P-consciousness exists (even though they use it to do the job!). Deciding that the robot has observed the natural world like we do is a serious mistake.

None of this argument declares that robot scientists are impossible. Indeed, hasn't this entire book come about because I am attempting to do exactly that? What it means is that in a future world, where a robot can do science, that robot must be capable of enacting every part of the chain of causality shown in Figure 8.3, and produce a novel law of nature, a novel statement in set T, through exposure to the previously unknown, uncharacterised natural world, like human scientists. Only then can that robot be claimed to have acted scientifically and proved it. Until then, if anyone says there's a robot scientist about, then you can scientifically test it by attempting to get it to do everything in Figure 8.3 on a completely novel aspect of the natural world, without human assistance. If it can't, it's not a scientist.

8.9 Alien science

In formalising the science process using set T to represent the collection of all explicit scientific statements, what can we say about the uniqueness of set T? Can there be more than one set? How could there be more than one? Aren't 'laws of nature' fixed, immutable, timeless rocks of knowledge? Clearly Kuhn would disagree, and has shown us that set T members are mental constructs regularly overturned and sent to set H. But this is beside the point. Remember, it was all done by humans, and it is the *contents* of a single set T that changed. Whole communities of humans wage critical wars to determine which statements get to stay in T and which are relegated to H. In the end, however, there is only one set T and one set H produced by human scientists. Call them T_{human} and H_{human}.

Nevertheless, and in principle, could there be another whole set T that has equal claim to be a set of scientific statements about the natural world? Based on the activities of human science one would say no, but we have no evidence/proof. The fact that we only have one T does not prove that our T is unique. We can use the framework itself to answer the question scientifically. There is no set T_{human} statement that proves there can only be one set T for our one natural world. We have already recognised that non-humans do not do science of the kind that populates a set T. But that lack does not demonstrate its impossibility. What if a bird could do science? Scientific statements by a bird would have a radically different form, but if translated into human-interpretable form, bird science outcomes might be

$$T_{bird} = \quad \{t_{bA}, t_{b1}, t_{b2}, \ldots, t_{bn}, \ldots t_{bN-1}, t_{bN}\} = \{empty\} \qquad (8.8)$$

$$H_{bird} = \quad \{h_{b1}, h_{b2}, \ldots, h_{bm}, \ldots h_{bM-1}, h_{bM}\} = \{empty\} \qquad (8.9)$$

Note I have emptied the sets. Birds solve problems and are intelligent, but they do not do a bird version of t_A/t_n human science. Yet in principle the emptiness of non-human sets shown in equation (8.8) and equation (8.9) is merely a practical side effect of the nature of bird mental capacities. In reality we have more than one instance of sets T and H, and the fact that one of them is empty does not prove that a second set is impossible. Indeed it looks quite natural to have more than one set. To get a new set T it appears all we need is a super-smart bird. Then, in principle, bird science could be as predictive as ours and yet have completely different contents that reflect the bird's cognitive capacities. Remember, all we ever claim about the statements in the set is that they make us predictive.

Take it further. Somewhere out there in the universe is an alien scientist who has P-consciousness that, say, directly perceives only gravitons and neutrinos. Such a bizarre P-consciousness would depict the same natural world totally differently. What we call matter, it might call space, and vice versa. The alien scientist, constructing its T_{alien}, would assemble statements equivalent to t_A/t_n in T_{alien}, abstractions −'laws of appearance'− that look totally different from T_{human}. Are they truly different? Each set, T_{alien} and T_{human}, was constructed using a unique P-consciousness, and that P-consciousness is implicitly built into the set T by virtue of the statements being predictive of appearances. So they are very different statements created as descriptions of the same natural world. Their equivalence is not in their appearance, but in their predictive utility.

Consider a simple experiment on Newtonian dynamics where human and alien are expected to predict the location of an accelerating apple at some point in time. Following their individual set T, both human and alien scientist point to the same outcome. They are equally predictive. Yet the abstractions in their respective set T might be as different as quantum electrodynamics is to Sanskrit poetry puffed in smoke signals. The science (of 'appearances') of alien and human are utterly different, yet both are identically predictive. It is the predictive utility of the statements (acting as laws of nature in some sense), that is the true invariant across all sets of the T kind. Remember: Set T is claimed to hold nothing more than a bunch of beliefs that, if internalised by a host 'believer', and if they are correct, will make their host predictive in a certain context. Nothing else is claimed.

Therefore there may, in principle, be as many sets T as there are kinds of creatures (a) with sufficient cognitive faculties that then (b) decide to behave as per a t_A of some kind, like we did. It is the difference between the mental faculties (in particular P-consciousness) of the creatures that will determine, to some extent, the differences between the set T of each kind of scientifically behaving creature.

8.10 Zombie science

Recent philosophical discussions have provided a useful way to encounter the most important aspect of t_A: the equation (8.4) "...*in principle can be refuted through the process of experiencing evidence of the regularity>*". The bolded words connect a measured physical phenomenon (scientific behaviour) to another scientific measurement paradigm: the 20 year old record of the scientific study of P-consciousness, a natural phenomenon that literally *is* scientific observation itself. To *experience evidence* is to use P-consciousness. That is the pivot around which the scientific world can turn.

To understand further how P-consciousness involves itself in a scientist we can make use of a vigorous discussion of last 20 years or so in philosophy: the philosophical zombie [Kirk, 2012]. This is a thought experiment. Philosophical zombies are not bloody, mangled philosophers sitting in armchairs arguing over their favourite XYZisms and who's brain to eat next. Philosophical zombies are exactly like us, atom for atom, but have no conscious experiences. P-consciousness is absent by definition. There is 'nothing it is like' to be a zombie. The life of the zombie is what humans get when in a coma, which is when everything but the brain stem shuts down. The difference with zombies is that they get to continue behaving normally whereas humans do not 'behave' at all in this state. Yet, according to the definition, zombies behave like us. By defining the zombie this way we are immediately confronted with the relationship between behaviour and P-consciousness.

We can make use of the zombie by seeing what happens if a zombie tries to do science. Apparently their behaviour is indistinguishable from ours. One human behaviour is scientific behaviour. Therefore the zombie should be able to do science. Another human behaviour is an ability to be confused about P-consciousness. Zombies are identical to us, so both of these things should be behaviours observable in zombies. Logical dissonance abounds in this. For example, how can this apparent zombie confusion about consciousness be genuine, when the subject material of the confusion is something the zombie has never had? Consider another human behaviour: death. Assuming zombies have to do that, too, the philosophical zombie could not notice it. Radical changes to P-consciousness happen in a coma state in humans, and the transition to death is when everything else in the body stops working in addition to the non-cranial central nervous system. Like the coma patient, there is nothing 'it was like' before death, so it makes no sense that a zombie could argue about the differences death makes to their existence.

Maybe there is a zombie afterlife, in which case the philosophical zombie's death would actually be a form of life! Personally I don't believe in afterlives. However, I do recommend considering the difference between the afterlife of the zombie (if there is such a thing) and the afterlife of the human. The transition, from the 1st person perspective of the previously awake, alert human to a deceased human, involves switching off P-consciousness. Death of a human then slowly disassembles all the atoms, so whatever process that caused P-consciousness before death has stopped. The zombie lost nothing. Therefore the philosophical zombie cannot possibly be identical to humans. This is a *reductio ad absurdum* style proof.

This is the nature of the gross contradictions riddling the definition of the philosophical zombie. On the one hand they have no experiences whatever. On the other they are somehow able to be confused about experiences that they have never had. They are supposed to be able to do science, which, we have measured, is critically dependent on experiencing radical novelty. Yet somehow that very natural world novelty is utterly unavailable to them. In a planet of zombies, all behaving like humans, not one of them is actually aware, in the way we are aware, of another zombie. There is no-one there to validate their behaviour as human because there is no experience going on. They operate 'in the dark'. No they don't, because 'in the dark' prescribes an experience: darkness. There is no darkness. All zombies have is what we have in an endless dreamless sleep. Somehow this world of zombies is a world of puppets with an invisible puppet-master that humans do not have, and who is responsible for all the knowledge underlying the zombie's behaviour.

Our analysis of scientific behaviour can help unpack this conundrum. Take another look at t_A. Now consider the idea of an "atom for atom copy". What are these 'atoms'? These are entities from a t_n describing the natural world of materials, arrived at through human scientific behaviour, the science of appearances that presupposes P-consciousness. Nothing in our depiction of the world of atoms would prescribe the existence of P-consciousness. Yet if we somehow got an elaborate 3D printer that could print out an atom for atom copy of, say, Albert Einstein in 1890, we'd have Einstein, right there with all his prior history and memories, using P-consciousness, who would then set about revolutionising physics. Have we printed out a zombie Einstein? What is missing from the printout that might absent P-consciousness from our printed Einstein? There is nothing in the scientific vocabulary (set T) that one can claim is missing. Yet whatever it is that generates P-consciousness tags along for the ride when the printer prints atoms. P-conscious Einstein then does science and proves he has P-consciousness by encountering the radical novelty of relativity, which culminated in its empirical verification and its subsequent depositing in the early 20th century set T. How would anyone argue that printout-Einstein is a zombie? This touches upon the very nerve of the whole issue: the apparent intangible immaterial, unknowability of the origins of P-consciousness. That missing understanding is literally the difference between the philosophical zombie and humans.

Instead of going on and on about the ineffability of P-consciousness and the recondite nature of its empirical investigation, why can't we just look at what the measurement of scientific behaviour (the t_A/t_n framework of this chapter) tells us in general about our scientific descriptions of the natural world? Maybe that can give us some insight into what the philosophical zombie is telling us. Firstly, note that no t_n predicts P-consciousness. They all predict the contents of an assumed P-consciousness. It is obvious that this should be the case because t_A tells us that P-consciousness (an ability to 'experience evidence') is presupposed and that statements constructed using it result in predictions that prescribe how the world will appear using P-consciousness. Of course observations cannot explain an ability to observe anything! It seems obvious. I am repeating myself, I know, but it seems necessary to get the concepts across.

Could someone have measured t_A/t_n say, 250 years ago? Maybe, but they would be unlikely to get an accurate t_A because they did not have the neuroscience and the language to clearly isolate all the elements of it (as per Chapter 3). If they did, however, the first thing they would predict is a long and ultimately intractable failure to scientifically account for an ability to scientifically observe anything. That intractability would then embed itself, as it has, for 250 years. Think about it this way: we have now recognised that we have at least 250 years of evidence of predictable intractability lending weight to the accuracy of the measured t_A/t_n framework.

Zombie science is revealing. Say a zombie scientist encounters a radically novel aspect of the natural world. That is the lot of the scientist: to encounter the unknown, distil regularity and put a statement in set T. By definition, novelty is something that no human has ever encountered, and when we encounter it we do so with our P-consciousness. By what magical mechanism is the behaviour of the zombie scientist, without P-consciousness, to be exactly the same as the human scientist? Logically then, the zombie must be *a priori* configured to encounter the novelty with behaviour identical to a human doing the same thing. But the zombie can't. The knowledge required to do that is what the zombie is trying to find! We've already said that the zombie is atom-for-atom identical to the human original. Its atomic composition is innately capable of producing P-consciousness, which we have denied in the zombie. Therefore the zombie cannot possibly react in the manner of the human scientist: the zombie must have already encountered the novelty in order that it behaves identically in respect of it. Therefore the encounter is not actually with novelty at all. Therefore the premise of the thought experiment must be false. The zombie cannot be atom-for-atom identical with the human.

8.11   The post-zombie science apocalypse

Getting back to the matter at hand, if, in a future world of a more mature science framework, scientists were to make a statement about the natural world that *did* predict P-consciousness in some routinely testable way, *what would that statement look like?* It cannot be of the kind that resulted from behaviour prescribed by t_A. That being the case, what is that future scientist actually doing? Whatever it is, it's not what we do now, nor is it anything that has been done for 2000 years of science. Is this future scientist, routinely making statements of an unknown nature, an impossibility? It's hard to argue a viable basis for its impossibility. The complete lack of these as-yet unknown statements does not prove their impossibility any more than a complete lack of observed black swans proves black swans are impossible. In that future world, I predict the scientists will look back at us, now, and wonder how we missed it all for 2000 years. They will look at the enormous pile of literature and be completely agog at how something that obvious was implicitly, culturally expunged to the point of systemic invisibility. It will be an historical artefact; an understood by-product of a process we had to go through, and that change wasn't possible until we were ready for it.

To help ready us for the change, let's imagine a world with a new kind of statement that does not presuppose P-consciousness. They are not like t_A/t_n. They do not go into set T, yet are scientific statements. One thing is obvious: these new statements, whatever they are, do not deliver P-consciousness itself. Just like existing statements, t_n, they are an account of the natural world, not the natural world itself.

Another thing that also becomes clear, as we face the view that the t_A/t_n framework presents of ourselves, is the level of disconnect between the natural world and our scientific account of it. The natural world itself, and our account, are two different things. Consider this: we say we humans are made of atoms. But we can't be. At least we can't claim to be according to what our science framework allows us to claim. Science statements such as t_n capture what the natural world *appears* to be made of, but only from the perspective of being inside that described natural world, observing it with the faculties resulting from that context. These are the claims you can make with the t_A/t_n framework. Nothing more. The obvious question, *"What is the natural world actually made of?"*, is not addressed. Acting as-if the material world is made of atoms makes us usefully predictive. That is all.

If our existing framework is claimed to literally address the actual structure of the natural world, then that claim must implicitly involve some kind of extra claim (about the relations of t_A/t_n with the natural world itself) that is not justified under the framework. One such extra claim might be *"The natural world is a great big computer running set T as a program"*. Such a statement would be automatically deposited in set H by the rigours of t_A/t_n. If you believe it or not, it changes nothing, predicts nothing new or testable and therefore has no right to set T.

But the natural world must be made of *something*. The t_A/t_n framework therefore tells us that whatever it is, it cannot be claimed to have been captured in set T. Secondly it tells us that whatever the natural world *is* made of, when assembled as a human you get an ability to scientifically observe because you have P-consciousness from the first-person perspective of being embedded within a massive collection of it, made of it. The conclusion is that whatever this new kind of scientific statement is, it is more closely related to whatever it is that the universe is made of, rather than what it appears to be made of. If such statements do not go into set T, where do they go? Are there examples of these new statements already discarded in set H and we just don't know it? What scientific behaviour is entailed in constructing and validating these new kinds of statements? What would the collection of all such statements look like? How do they relate to the existing set T? Does the new kind of statement invalidate the old, or do they live side-by-side with them? How do they get empirically validated?

8.12 Letting it settle in

Please take time to internalise the above framework: it was determined *empirically* by observing the natural world. The only difference is that the natural world observed is us, *scientists*. Scientific argumentation by empirical testing is an argument settled by measurement. Contest the framework, but leave out any/all arguments that have no measurable basis. To perform the measurements we turned to behaving scientists, not literature. The literature is made of scientists/philosophers writing what they *think* scientists do, not measuring what they actually do. Its concern mixes up the product of scientific behaviour, its practical logistics/methods and the scientific behaviour itself. Setting this literature aside has meant, at a minimum, setting aside the entire suite of literature produced by the philosophy of science. It's very interesting, but if you read it or not, it changes nothing in the framework. Scientists are not required to read philosophy of science, and most never do. Scientists sometimes write about science (instead of their own output), but most never do. We simply don't need it, for the same reasons we never needed it to characterise any other aspect of the natural world and be successful.

8.13 Summary

We never ask lab animals what they think about their behaviour in the experiments they are in because we know it to be irrelevant data even if they could answer. The somewhat grounding reality for us scientists is that this time we are the lab animals. In this chapter we detailed, by measurement of scientific behaviour, the t_A statement in the previously constructed $t_A/t_n/T/H$ framework. In statement t_A we see the fundamental dependence scientists have upon the neurological faculty P-consciousness for acquisition of all scientific evidence. We call it an act of scientific observation. We are indeed doing what history tells us is 'organising appearances'. This reinforces the previously revealed anomalous state of the science of consciousness: we are using 'experiencing evidence' to 'explain' our faculty for 'experiencing evidence'. In that paradoxical circumstance all we get is 'the appearance of the experience delivery process', not why it happens. As an account of P-consciousness it has the form 'animals run by running'. The logical fallacy, a product of a scientific evidence boundary condition at the heart of this, is now as clear as it gets.

In this chapter we also encountered the possibility of a new, completely different scientific behaviour that results in a different form of scientific statement. It pre-warns that such statements are about the natural world itself, as opposed to what it appears to be to us scientists, the presupposed observers. The existing $t_A/t_n/T/H$ framework is a product of the mind. The new kind of statement describes the universe in a different way, prior to the existence of an observer, but obviously it must be consistent with what we observe while at the same time accounting for an ability to scientifically observe at all. The new set cannot be a product of a mind, by definition. Can you see a way such a new statement can be produced? That is covered in Chapter 11.

To proceed from this point note the $t_A/t_n/T/H$ framework is revealed to depend on an ability to produce statements intelligible by fellow creatures of the same or similar cognitive kind. To go further we need to do the neuroscience of statement formation (belief dynamics). Not only will it make the basic science framework a tangible output of brain tissue, it also sheds light on more generalised learning such as the cultural learning that we have already covered. It is to that neuroscience we now turn.

Faced with an admittedly fundamental anomaly in theory, the scientist's first effort will often be to isolate it more precisely and to give it structure. Though now aware that they cannot be quite right, he will push the normal rules of science harder than ever to see, in the area of difficulty, just where and how far they can be made to work. Simultaneously he will seek for ways of magnifying the breakdown, of making it more striking and perhaps also more suggestive than it had been when displayed in experiments the outcome of which was thought to be known in advance.

[Kuhn and Hacking, 2012, Page 87]

Chapter 9

The Biology of Belief: Statement Formation

Still, to say that resistance is inevitable and legitimate, that paradigm change cannot be justified by proof, is not to say that no arguments are relevant or that scientists cannot be persuaded to change their minds. Though a generation is sometimes required to effect the change, scientific communities have again and again been converted to new paradigms. Furthermore, these conversions occur not despite the fact that scientists are human but because they are.

[Kuhn and Hacking, 2012, Page 151]

In this chapter, in addition to complex dynamical systems ideas and neuroscience knowledge, I am going to use all the concepts covered in the previous chapters. With these I will construct a modern picture of the human capacity to make statements (behave), to learn to make novel statements, and to inhibit/alter existing statements. These concepts will be directly brought to bear on our ability to do science and it implications for changes to science itself. The material in this chapter presents standard concepts in a novel way. For background on the basics of dynamical systems theory please use [Izhikevich, 2007] and [Shapiro, 2013] as a both good technical references and a good launching place into the rest of the relevant literature.

9.1 Dynamical systems

The human brain is probably the most complex single entity known. Increasingly, the most powerful descriptive methods applied to brain function exist within dynamical systems and complex systems theory. Combined with the relevant neuroscience, this approach treats the brain as a complex object (in our case 100,000,000,000 neurons and

100,000,000,000,000 synapses linking them together) that has a complex dynamical 'state' that, in principle, is simply a large number of simple states all interacting with each other. The idea of 'state' can be a little confusing at first. We begin by ensuring we have a basic grip on what it means.

A system's state is captured by one or more properties that completely specify the target system. Consider Figure 9.1, where we see

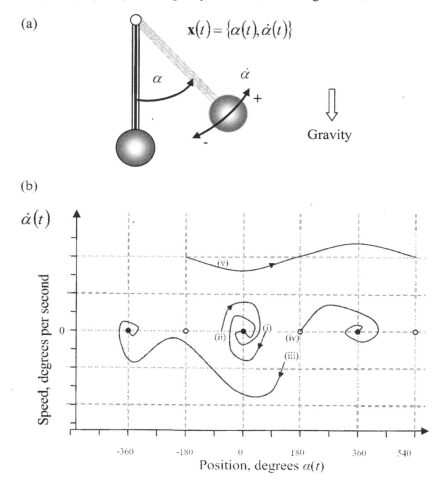

Figure 9.1.   Dynamical systems basics: the un-driven frictionally damped pendulum. (a) The state vector **x** lists individual cases of valid combinations of speed and position for the system. (b) State trajectory diagrams for the system.

an undriven pendulum in a uniform gravitational field (operates in a flat plane like a grandfather clock). The pendulum has a mass on the end of a rod. The pivot point is imperfect, and friction slows the pendulum. To completely specify the state of the pendulum requires the pendulum's position α and its speed $\dot{\alpha}$, both of which vary with time t. Position and a speed are the 'state variables'. A list of a particular instance of these numbers is called the 'state vector' (shown as **x**). The number of state variables is called the 'dimensionality' of the system, which is why we say the dimensionality of the state vector **x** is two. The more complex the system, the more state variables are needed to uniquely specify the system. The full potential range of the values of the state variables defines a 'state space' or 'phase space'. The state dimensionality of the pendulum is two. We can therefore also say that its phase space has a dimensionality of two.

Because the dimensionality is only two, we can draw a state trajectory diagram on a sheet of paper. This is shown in Figure 9.1 for various pendulum trajectories based on an initial position and an initial speed. Trajectories (i) and (ii) show the pendulum released from two positions to the left/right of $0°$, and less than $90°$. The decaying sway of the pendulum to rest makes a spiral to the origin in phase space. In (iii), if the pendulum is raised nearly to the top right and then pushed hard clockwise, it goes around more than a full rotation and comes to rest at -$360°$, which is pointing down as before. In (iv) the pendulum is raised so it is upside down and balanced. The slightest perturbation causes it to lose balance, in this case counterclockwise, oscillating to rest at +$360°$ which is also straight down. Finally, in (v) the pendulum was raised clockwise to be vertical again, but this time it is shoved so hard counterclockwise that it goes all the way around and keeps going off the graph. With variable mass, and friction and the length of the rod, the 'phase portrait' allows complex trajectories, even for this simple system with dimensionality two. In all the trajectories shown in Figure 9.1, the start of the trajectory is when time $t = 0$.

Note that we do not have to elaborate the equations for the system to understand the system at this level. When you write down the equations and analyse their behaviour for different initial conditions, you find that there are special points in phase space called equilibria. These are shown in Figure 9.1 with • for a stable equilibrium point (straight down pendulum), and ○ for an unstable equilibrium point (straight up pendulum). Equilibrium points are very important determinants of the behaviour of the system. Stable equilibria are called 'attractors', and the trajectory tends to bend towards them. Unstable equilibria tend to cause the trajectory to veer away. Their influence is usually called a bifurcation point where the trajectory goes on to one of several paths, but it is very difficult to know which path will be taken, and each can be radically different in nature. If there are more than two state variables (dimensionality greater than 2) then the trajectories can be vastly more complex, and more difficult to draw and visualise. Their stable and unstable equilibrium points can cause highly unpredictable results because around the point, tiny differences in trajectory can cause dramatic alterations to the behaviour of the system.

9.2 Non-stationary systems

Imagine there was a pendulum whose mass was able to change, or whose rod length was variable or whose pivot friction changed. In that case the equations for the system are changing, and the dynamics of the changes introduce their own dynamics into the system. The phase portrait is like a contour map, where the contours are moving around, morphing into different landscapes on the fly. This is a non-stationary system. However, it is still not complex enough for application to brains. Imagine, now, that new pendulums can arrive, each with its own dynamics. Each of these additions adds to the state space dimensionality. The number of states increases more dramatically with each new pendulum.

Now imagine that each pendulum is variably linked to an arbitrary number of other pendulums by connections that can vary in strength. Each pendulum 'pushes' a large number of other pendulums. The net effect of many pendulums is to push a single pendulum. The phase trajectory landscape is now so multidimensional and changing so rapidly that it's hard to imagine it diagrammatically. But there is one way to understand it: *you are one of these systems*. There are a huge number of 'pendulums' in each neuron. Each of these is linked in a certain way. Then the entire collection of 'neural pendulum collections', with a dynamics of its own, is variably linked to many other 'neural pendulum collections'. Links to other neurons involves two very different mechanisms. First via electrochemical signalling called action potentials (AP). Second via electromagnetic (EM) coupling similar to a short-range radio except there is no real broadcasting. The 'pendulum' linkages due to AP and EM link cohorts of pendulums in entirely different groupings. If there are N (in the thousands, say) pendulums for a neuron, and 100,000,000,000 neurons,[a] each linked to 10000 others via AP and 100,000 via EM, and all connections are varying in strength on a moment to moment basis, then you start to appreciate the staggering complexity of the organ between your ears. We can't ever draw the phase portrait for such a high-dimensional system. However, we can use a lower-dimensional phase portrait to get some appreciation of it. This is shown in Figure 9.2.

No matter how complex the system behaviour, and the degree of parallelism, its trajectory in phase space is a single point taking a complex path through a dynamic landscape including a multitude of variably stable and unstable equilibria. In Figure 9.2 this complexity is shown over an extended period of time. Gradually a picture appears of the gross, long-term behaviour of the system. No matter how complex the path, there are some regions in phase space that are never on a trajectory. Limits to the variability of the system become apparent. This island of state-space represents the entire suite of behaviours for the entity hosting it. The island has holes shown in Figure 9.2(d) regions R_2. The island has an outer shore delineating another 'hole' labelled R_3. We may pay attention to a particular region such as Figure 9.2(b) R_1, and it may be associated with a particular 'statement' of interest. Notice that the Figure 9.2(d) form has completely lost the sense of particular trajectories. It only indicates allowable zones of state space.

---

[a] In some regions neuron populations also vary because new cells are created from neural stem cells (neural genesis), and existing cells die (neural apoptosis). This too, is a mechanism supporting dynamics of a particular kind that adds to the diversity of the state trajectory.

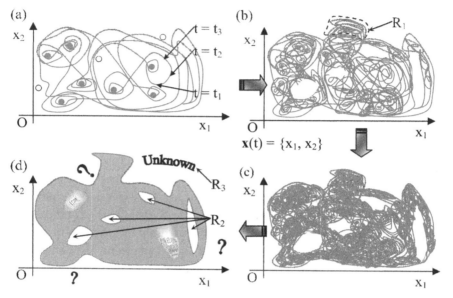

Figure 9.2.   A complex trajectory landscape. (a)...(d) shows a complex trajectory involving dynamically appearing and disappearing stable and unstable equilibrium points. As time progresses, the state trajectory reveals a particular shape characterised by 'no-go' zones.

9.3 Pendulum 'statements' in phase space, and its generalisation

Consider again the simple pendulum. Imagine its *trajectory in phase space to literally be the pendulum making a 'statement' in the real world*. This basic idea for a simple pendulum is the key to understanding how high dimensional systems like the brain operate. No matter how high the dimensionality, and how variable the state trajectory landscape, it is still a basic reality that the state trajectory is directly associated with some kind of physical behaviour in the real world, even if the system is an alert, awake human, and even if the path is not particularly predictable or unique from time to time. None of these things matters in principle, because despite the huge dimensionality we are able to consider the output of the brain, as a determinant of physical behaviour of the body it inhabits, as a 'statement' made by that body. One statement may be a serve in a game of tennis. Another statement may be to verbalise t_A. For an embodied human cognitive agent, this is meant literally, and physically, and is directly analogous to the pendulum 'statement'. Each statement involves the state of the nervous system, as a whole, progressing along a specific path. The fact that human behaviour is more complex than the pendulum is purely a matter of degree, not of basic kind.

Next consider two human cognitive agents, with a state space identical to Figure 9.2(d), who may be in exactly the same state (location) but have different behaviour because their trajectory (direction) is different. Each makes a different statement. Now consider two cognitive agents, again with identical Figure 9.2(d) trajectory maps, who make the same statement, but their individual state trajectories may be very different. How do we compare and contrast two cognitive agents? The long tradition of behaviourism, which lasted from early in the 20th century roughly until the mid 1960s, would have it that the external behaviour uniquely reveals internal state. We now know that to be simply false. Not only can the same behaviour not result from the same state, it need not even result from the same trajectory in the same individual from one moment to the next. There are many ways to skin a cat, so the rather disturbing saying goes.

How do we compare and contrast the state trajectory behind the 'statements' of two cognitive agents?

Evolution solved that problem a long time ago: *you don't*. The only function that has to remain simultaneously adaptive (learned) and yet repeatable (invariant) is external behaviour. Just like my experience of red does not have to be yours for us both to report red and behave identically in response to redness, my internal state island and trajectory does not have to be identical to yours to produce the same statement (such as a report of redness). Evolution chose this so that you can avoid the hungry tiger. You do it or you die. Those that couldn't are out of the gene pool. That avoidance is a kind of 'statement' like tennis playing is a statement. It is a bodily behaviour in an environment. Evolution does not care *how* you do it, just that you must do it.

The obvious generic, abstract version of this evolutionary imperative is '*you develop appropriate statements or you die*'. The abstraction we use here is the act of creation and delivery of statements. This completely turns the behaviourist paradigm around by asking what kind of system can consistently adapt (*change*) to produce the same statements despite very different environmental conditions, immediate stimuli, bodily condition, developmental condition, genetic makeup and other happenstance. All these things combine to create an individual's state trajectory such as that confined within the Figure 9.2(d) 'island of statements' at a moment in time, complete with all its go and no-go zones. As long as the individual's state landscape can produce the same physical statements, then that individual can adapt to and exist in an environment that can be radically changeable yet express deep regularity for which statements form a viable, useful goal.

Figure 9.2(d) also teaches us the difference between possible and impossible statements (behaviours). Impossible statements are outside the state-space landscape, inhabiting the regions marked 'unknown'. This region is beyond the present capability of the host. Knowledge of quantum mechanics, Chinese language or an ability to juggle might be in this region. There may be a region within the 'possible' that is unexplored. This might be the result of having a skill set capable of solving a particular problem, but having never solved that problem. Part of education is to prime new regions of state trajectory so that encounters with certain classes of novelty yield to our efforts. Without the basic skill set, however, the problem will remain intractable (in the 'unknown' region of Figure 9.2(d)).

State-trajectory landscape dynamics has a very simple way of explaining the Kuhnian notions of normal (paradigmatic) science as puzzle solving and the abnormal science involved in paradigm shift. In the former, the paradigm lays down the infrastructure for the scientific puzzle solving (called 'normal science' by Kuhn). It establishes, on a community level, the boundaries of the 'unknown' meant in the sense of Figure 9.2(d) for each member of the community. The fine details of particular trajectories, that are the solutions to puzzles within the paradigmatic 'state-space region', may resist efforts for quite some time, but ultimately someone will express/traverse the trajectory of the correct solution. In contrast, when a paradigm's state space region does not hold the solution, what is encountered is a serious anomaly that renders the paradigm a defective approach. The actual solution is in the unknown region Figure 9.2(d). It is simply impossible for the paradigm to 'utter' the solution to what will then become the crisis-provoking problem. The paradigm itself has to change, which means a whole new region of state-space landscape must be constructed, within which lies the solution to the anomaly and, eventually, a new class of as yet unknown problems. The state-space approach lends great simplicity to the processes described in Kuhn's *Structure*.

In this there is a subtlety inherent in the idea of novelty. Novelty can be a new trajectory through an existing state landscape. Novelty can also involve the creation of a whole new region of state trajectory landscape. In both cases, the existing well-travelled trajectory is directed towards new terrain. It is a matter of degree. To get to a whole new state trajectory region requires the normal, well-travelled trajectory to be significantly diverted. In dynamical systems terms, this means that a novel attractor must be created to substantially alter the landscape to divert the trajectory towards a new region. Clearly, a less significant 'jolt' to the trajectory is required to handle less significant trajectory shifts. A whole new region of trajectories requires a more substantial 'jolt'. Later in this chapter I will more formally define these kinds of alterations to trajectory, and calibrate them in more familiar terms.

9.4 Nonsense statements

"*Twas bryllyg, and ye slythy toves*" begins Lewis Carroll's poem Jabberwocky. This is a nonsense statement produced by Carroll that came from the state trajectory of his brain tissue as the originator of all the muscular activity behind writing the nonsense. The statement is unusual in that its content is the deliberate absence of content. As English speakers with knowledge of the symbols of the standard English alphabet, you and I can imitate the act of uttering that statement, and quite accurately say nothing sensible yet convey the nonsense itself to the host of another receptive state trajectory in another cognitive agent like ourselves. Consider the state trajectory involved in the statement

"*One plus one equals frog*" in Figure 9.3(a). My personal state trajectory was able to create that statement. Unlike Carroll's statement, "*One plus one equals frog*" is at least recognisable as a syntactically accurate sentence of modern English. What is remarkable is that these statements are, at the moment of their creation, the result of a completely novel state trajectory. Yet the adaptation and dynamics involved in its production is a natural and navigable part of the state trajectory landscape of its originator (me).

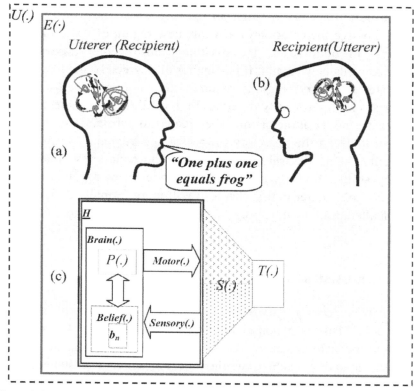

Figure 9.3.   Within universe $U(\cdot)$, is environment $E(\cdot)$ in which is embedded two cognitive agents, (a) an utterer of a statement and (b) a recipient of it. Both (a) and (b) have similar perceptual and sensory/motor systems of kind (c), and each agent's state trajectory is totally different. What has been conveyed from (a) to (b), the statement, is a *token* that represents the state trajectory of (a), and that 'looks up' or 'indexes into' the state trajectory of (b). At the instant of communication, this process has nothing to do with the meaning of the statement.

Consider the particular portion of the state trajectory responsible for the utterance of the word 'frog' in Figure 9.4(a). If I now alter it slightly I get *"One plus one equals two"*. The difference is shown in state trajectory terms in Figure 9.4(b). At one level the statement is indistinguishable from Figure 9.4(a). To a cognitive agency whose phase space includes an ability to understand English, each is a syntactically, grammatically accurate statement of English. Both statements would sound like Jabberwocky to a non-English speaker. A parrot could be trained to utter both statements. However, if the utterer of the statement Figure 9.3(b) has an appropriate developmental/learning history, then the second statement represents a mathematical truth in a certain class of number theory. Uttered statements consistent with the abstractions of number theory can very usefully cohere with the state trajectory of a similarly trained recipient of the statement. That coherence, however, is only incidentally transported via the 'statement' that mediates it, and is entirely dependent on the capacities of the recipient's state trajectory. This is the manner in which I intend all such transactions between an agent with a nervous system, and the environment containing that agent.

The fact that there is no obvious context in which a particular statement makes sense does not entail any sort of fault in the trajectory

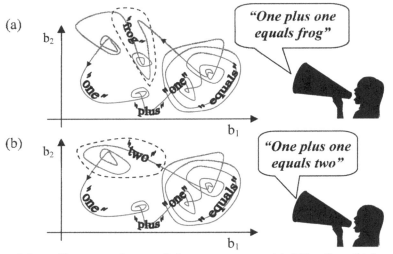

Figure 9.4   The state trajectory of abstract statements. (a) differs from (b) in a way visible in the differing state trajectories.

landscape of the utterer. It merely makes the statements of little value as communication. My brain state trajectory might accidentally utter the exact solution to a great mathematical conundrum without ever actually knowing it. A pre-linguistic human, a million years ago, might be quite able to be trained to say $E = mc^2$ without having any clue as to its meaning to 21st century physicist's brain state trajectory.

This is the way that it makes sense to characterise 'statements' made through the traversal of a complex state landscape. The statement gets its context from the utterer's state trajectory, and a receiver's state landscape, guided by the tokens in the received statement (auditory phonemes or written/visually conveyed syllables or other signalling), that elicits some kind of future behaviours (state trajectories) that may or may not have anything whatever to do with the state trajectory (as represented by the word 'intention') of the original utterer. We need attribute no accurate semantics to any portion of this entire transaction. The statement itself should be regarded as complex token-generation resulting from the state trajectory of the utterer. On receipt it becomes a kind of 'look-up token' for the receiver's state trajectory to respond to. It is the extent to which the utterer's and the receiver's state landscapes include a level of coherence about the meaning of the tokens, as reflected in behaviour, that determines what has been communicated between utterer and utteree.

Previous chapters, along with these considerations of nonsense statements have led us to a position summarised in Figure 9.3 where we see two cognitive agents (a) and (b) of similar kind (c) whose internal, and fundamentally dissimilar state trajectories are modified during the process of (a) uttering a statement and of (b) receiving it. As an instance of cognitive agency of the kind shown in Figure 9.3(c), our utterer (a) interacts through S(.) with a T(.) that is literally cognitive agency (b), the recipient. For the recipient, however, T(.) is the utterer. This is a reciprocal relationship mediated by the natural world S(.), which is literally identical for both Figure 9.3(a) and Figure 9.3(b). In this case, if the statement is made verbally, S(.) is the air mass transporting the sounds of the statement along with any olfactory chemistry. S(.) may also include the space carrying the light between the two. All modes of available P-consciousness P(.) in (a) and (b) are thereby impacted, all mediated by the shared $S(.)$. All of it impacts the state trajectory landscape of utterer and recipient.

In this way, our discussion has disconnected all meaning in the statement from the behaviour that generated it, which is entirely a result of the history, capacities and motives of the utterer as contained in the state landscape of Figure 9.3(a) as a cognitive agent. In the recipient, the inverse process elicits a state trajectory of some kind, and once again we demonstrate that the meaning of the statement need only be thought of as a 'lookup index' into the state trajectory of the recipient. What happens after that depends entirely on the history, capacities and motives of the recipient as contained in the state landscape of Figure 9.3(b) as a cognitive agent.

This process is essentially identical for all creatures that have a nervous system. It is only different in the implementation details, in the operational scope of the various parts of their Figure 9.3(c) agency, and the sophistication of the state trajectory landscape of each (complexity of the nervous system). Figure 9.3(c) could equally well be used to elaborate the relationship between a mollusc and the rock it is sitting on. This is the intended level of generality and forms our base level understanding of the kind of cognitive agency that results in all behaviour (statements), and within which is contained all the essential components of scientific behaviour and the production of scientific statements.

> "Do words and thoughts follow formal rules?" One major thrust of the book has been to point out the many-leveledness of the mind/brain, and I have tried to show why the ultimate answer to the question is, "Yes – provided that you go down to the lowest level – the hardware – to find the rules.
>
> Douglas R. Hofstadter[Hofstadter, 1980, Page 686]

In the overall theme of addressing scientific behaviour, I hope this discussion has completely dissolved all tendencies to regard scientific statements as being in any way special, except as subsets of the intended general class of all possible statements, which for humans can include tennis playing, building a new scientific instrument, writing a nonsense poem and elaborating a new 'law of nature'. At this point, if you believe scientific statements are different or special or unique, or not, it makes no difference to the mechanisms underlying their creation, change and disposal, which is 100% cognitive/neural.

9.5 More on the dynamics of changes in dynamics

Figure 9.5 depicts the radical remodelling of the state trajectory landscape that can result from exposure to the demands of a sufficiently novel problem. Until the landscape including the solution is present in sufficient detail, the solution will be unapproachable. The event that may trigger access to the final solution can be anything. We all have stories of how 'the idea' happened in the shower or when stepping on the bus. This is the actual mechanism responsible for that event. There has to be a first-time through a new trajectory, and the event can be quite a striking experience.

9.6 Associative memory and the state trajectory

All of the dynamic processes depicted so far are a representation of a brain. When a brain operates in an act of recall we are accessing memory. This is the kind of memory system/P-consciousness associative resonance depicted in Figure 7.2 in the chapter on cultural learning theory. That associative resonance process is captured by a state trajectory traversal. When something new is stored, we establish a new state trajectory. And so forth. Trajectories associate. They may be nearby. They may join one resonance to another. When one trajectory leads to another then we get a form of reasoning called associative memory. When we think of uncles, then all nearby uncle-related trajectories become activated. If we think of pains, then all pain related trajectories become activated. If event A is repeatedly experienced associated with another event B, then A is added to the set of all B-related trajectories. All kinds of memory systems and learning (as examined in Chapter 7) can be re-phrased in state-dynamics terms.

9.7 Induction – problem solved (again)

Encoded in the previous sections is a solution to the problem of induction, which has a centuries-old history. It is a problem of reasoning intimately enmeshed with scientific behaviour. It occurs when we take specifics and leap to the general case. It occurs when we presuppose that the future will unfold as it has done in the past. In these examples of mental gymnastics, we are, in fact, making a formal error in logic. Yet we do it all the time and it works. In being able to make an error like this, we also gain access to new ways of thinking that may be quite accurate. This is especially relevant in science, where, in effect, we have to make mistakes to make progress. Whenever we formulate a hypothesis we are formally, purposefully making a mistake. We are making a statement that is not yet a scientific statement. We then set about challenging the 'mistake' by confrontation with nature. If it passes the test then it

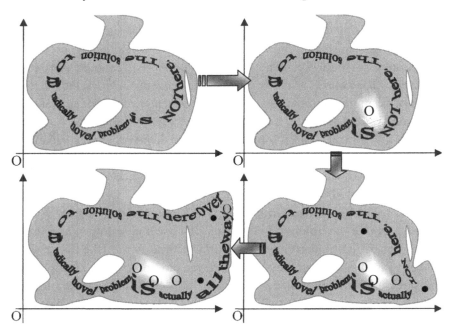

Figure 9.5.   The re-landscaping of a state space resulting from confrontation with radical novelty. A problem cannot be solved until the available trajectory space includes the solution to the problem. The changes result from repeated exposure to stimuli and behaviours that are represented by the stable and unstable attractors ● and ○ respectively.

becomes scientific and we keep it. If not, we set it aside as described at length elsewhere here.

In the scheme of the scientific targets addressed in this book, the problem of induction arises because we have no access to the underlying causality that results in the natural world unfolding the way it does. Because we are limited to appearances, we assume appearances will generalise from special cases (e.g. apply a broad classification to an individual object), and that the future will repeat the past. That, after all, is the nature of regularity. We do this all the time in science. If we had descriptions that captured the essence of the underlying reality (causality) then reasoning with those descriptions would not be inductive. It would be deductive.

How is the problem of induction manifest in state-dynamics terms? The neuroscience and dynamical systems approach makes short work of the problem of induction. Indeed I have already demonstrated it in Figure 9.3. Any old concatenation of trajectories will do. If you append trajectory A on to trajectory B (as per the above associative memory example) then you have connected one thing with another in precisely the manner of induction. The classic example is *"I have only ever observed white swans"* might become *"All swans are white"*. Indeed, the whole of brain operation is designed to permanently make the mistake of induction. It s tht smple. t's wht yr memry sstms do al t tme! And yet, despite this logical fallacy operating at the heart of the mechanics, science works really well in the sense of its capacity to be predictive. In more general human behaviours, clearly the brain's inductive capacities work or we wouldn't be here to be confused about it.

The problem of induction is therefore a non-problem caused by an expectation that externalised statements of appearances, no matter how formal, inherit the underlying logic of the natural world (the underlying causality) in them, when they do not, never did and cannot in principle. The human behaviour called science operates by being serendipitously 'not wrong' in the sense of being predictive. It does not do anything that can ever be claimed 'right', whatever that means.

9.8 The mathematics of statement dynamics as belief dynamics

State trajectory traversals that result in statements can be thought of (with your own state trajectory!) as reflecting the existence of a 'belief' in a very general sense. A belief might be manifest in the tacit knowledge underlying the 'statements' that are a game of tennis. A belief might be manifest in an ability to turn tacit knowledge of quantum mechanics into a concise mathematical statement of some kind, and convey its applicability to the natural world. The state trajectory concept tells us that there is no specific place in us where a belief is stored. The act of recall is a state trajectory and the traversal of the state trajectory itself literally is the act of statement of the belief. An act of remembering or learning modifies a state trajectory in some way so that its later traversal includes the modifications. There is no sense in which a belief need be located in anything else or anywhere else other than a state trajectory. The process of creation of a new belief or the modification of an existing belief is all about altering the natural dynamics of state trajectory traversal in such a way that the preferred belief is what is expressed when it is interrogated by the appropriate stimuli. A very long process exactly like this is literally writing this book and has just reached the end of this sentence. My particular statements are written, not spoken. It is hoped they elicit a response in your personal state trajectory, some mental imagery that reflects my 'belief' in respect of how beliefs are formed, changed and uttered.

One could dig out a whole pile of words like 'belief' and examine them for their correlation with the state-trajectory concepts of this chapter. We have already come across such words as 'understanding', 'knowledge', 'fact', 'truth' and 'theory'. In the context of this articulation of statement formation and conveyance, none of these words help because they operate at a layer of description above that of the basic biology of cognition. All we really need is to abstract the word 'statement' in a useful way, and let complex system dynamics and neuroscience do the rest. This is why I have steadfastly confined us to the idea of 'statement' and utterance throughout this book. I have paid attention to the word 'belief' because it has value in future discussions of scientific behaviour, where the production of highly calibrated, specific beliefs, deposited as 'statements', are a mandated, intrinsic part of the behaviour.

The dynamics of statement formation and change can also benefit from some of the mathematical concepts within dynamical systems theory. We know that a system, no matter how complex, has a state vector $\mathbf{x}(\cdot)$. In the case of a brain, the state vector has a very high dimensionality (number of state variables) and that each one varies with time. We know that the state vector's content, plotted out as a coordinate in a very-high-dimensional graph that varies over time, reveals the state trajectory of the system. We also now understand that not only does the system dimensionality itself alter over time (e.g. new neurons, neuron death), but that the individual existing elements of $\mathbf{x}$ change (e.g. new synapses, synapse loss, synapse weight). Therefore we are entitled to ask how we might regard $\mathbf{x}(\cdot)$ as containing or representing the process of change in trajectory. Not only is the exact state of the system contained in $\mathbf{x}(\cdot)$, this vector also contains within it exactly where the trajectory is headed, which intrinsically and automatically invokes the processes of change. Built into $\mathbf{x}(\cdot)$ is how some state variables stop contributing and new ones arise. New trajectories arise, old ones die out or fade to irrelevance. This is the process of change we call adaptation or learning.

Change, in mathematics, is formally captured by the process of differentiation. This is about as mathematical as I will get in this book. In the case of our time t dependent state variable,

$$\mathbf{x}(t) \tag{9.1}$$

the first derivative (differentiation) with respect to time is

$$\frac{d\mathbf{x}(t)}{dt} = \dot{\mathbf{x}}(t) \tag{9.2}$$

and the second derivative is

$$\frac{d}{dt}\left[\frac{d\mathbf{x}(t)}{dt}\right] = \frac{d^2\mathbf{x}(t)}{dt^2} = \ddot{\mathbf{x}}(t) \tag{9.3}$$

You don't have to fully comprehend differentiation of something to understand that it invokes some kind of conversion that formally reveals how changes over time are encoded within that something. You can use the same conversion again to reveal changes in the changes. And again to reveal changes in the changes in the changes, and so forth.

In our special context, the first-derivative has an interesting interpretation. While **x** is traversing the state trajectory corresponding to the expressing of a certain statement, the first derivative tells us how the state vector is changing with time. It contains a level of degree (speed or rapidity) and a level of direction. Equation (9.2) can therefore be thought of as analogous to 'velocity', in contrast to $\mathbf{x}(t)$ as specifying a position. In this way, there is a built-in mechanism within each element of the state vector that specifies some level self-alteration, so that the next time a particular state is reached, it can go in a new direction. In the intricate detail of each vector element of **x**, if there are local/cross-interdependencies related to the magnitude of the changes, the next time the exact same state (position) is reached from the same direction in state space, the state vector may go in a different direction. The statement is altered, or perhaps a whole new statement will now be produced. What is this if not an abstraction of the process of adaptation required to carry out a conversation, or write something? Furthermore, depending on the timescales and level of repetition of state trajectory traversals, and the adaptive properties of each state vector element, the new trajectory may become relatively permanent, like wheel-ruts in a road. This is the mechanism we call 'habit' or instinct or other well established behaviours and skills.

An even more interesting interpretation of the 1st derivative is that it is somewhat related to the concept we label 'reasoning'. If we follow a state trajectory (such as the one writing this sentence) from one memory association to another and then another via all the innate linkages of association, then what we are doing is performing a kind of reasoning. Like the act of reasoning that created that very sentence or the nonsense sentences created elsewhere here. Every one of these sentences is novel, for me. Every one of these sentences is coming from my brain state trajectory and its moment to moment diversions from the current word to the next words of the sentence, controlling my fingers as it goes. This is the action of the first derivative.

Consider what might happen if the first derivative were very low in some localised but critical part of the brain happen if the first derivative were low but maybe regionalising in some critical part of the brain first derivative were low in some localised but regionalising critical part of the brain zero in some localised but critical part of the brain localised but critical part of the brain critical part of the brain critical part critical critical critical. Have you ever met someone that can never quite finish a....

If the first derivative is akin to velocity, then the second is akin to acceleration – change in velocity, which we all know relates to force. How did I mentally force the previous paragraph to occur? I imagined what might happen if I mentally turned down the acceleration (the second derivative) in a critical part of the brain. As a result the velocity in that region may become constant and limit the range of trajectory in a language centre. I then wondered what might happen if the velocity slowly approached zero. I think you get the idea.

What about the second derivative more generally? Like a child's swing, the brain is a dissipative system, and needs some kind of repeated jolt to retrigger continued activity. The second derivative seems to connect with that aspect of brain function. What if the second derivative were larger and a bit randomised? Only geeks stuck in the 90s still go for dot-com organisational alignment and 'Outside the box' asset innovation. Perhaps Finnegan's wake is infinity's sleep and James Joyce was actually a physicist exploring observer-book entanglement. Really? I used to play cricket. Word extinction something much more random could have played sound to the ear, the ear goes to the sound. When you blot out sound and sense, what do you understand? While listening with ears one never can understand. To understand intimately one should see sound.

Well guided, the 2nd derivative can be seen as the power house behind innovation. In Figure 9.5, how is it that the state trajectory is perturbed into entirely new zones of phase space, for which there is no precedent? The only way an excursion into new territory can occur is through persistent encounter(s) with sufficiently radical novelty. So the second derivative is related to some measure of a capacity to handle novelty. If my 2nd derivative is so small as to render me incapable of learning about something novel, then what aspect of my being has been inhibited? It starts to look like the 2nd derivative is loosely related to intelligence.

This is only the briefest of skirmishes with velocity and acceleration, and I only claim to have shown there is something to examine, not that I have accurately captured the properties 'reasoning' and 'intelligence'.

There's another word that needs a little attention: rationality. I like to think that rationality operated throughout this entire chapter. Yet throughout the chapter, all manner of absurdity was rationally produced. This is quite paradoxical. The answer is clearly neurological at its base, and I find no better account of it than that of one of Karl Popper's students:

> ... neither beliefs nor acts of belief, nor decisions, nor even preferences, are reasonable or rational except in the sense that they are reached by procedures or methods that are reasonable or rational. (The phrase 'rational belief' is rather like the phrase 'fast food'.) Still less are beliefs, or decisions, or preferences ever justified, even in the weariest sense of the term. But my thesis does not imply that people cannot think or make decisions reasonably or rationally. It locates rationality in the way that people think not in what they think.
>
> David Miller [Miller, 2005, Chapter 5]

There clearly is a huge rational process operating in cognition: the basics of neurology and the natural regularities that make it happen. The natural world stimulates me with its own regularities (even persistent total randomness is a form of regularity), and I inherit that degree of 'rationality' simply by being in the world, and being awake, alert and healthy. If I disconnect myself from the world by taking psychoactive drugs, then the natural regularities in my brain are not as well connected to my sensory apparatus, and I may be irrational from the point of view of another agency like myself. Yet the natural world that is my brain is only doing what it naturally does when affected by abnormal natural circumstances. The natural world is still being 100% rational, but I, being drug addled and blathering about the cats in my fridge, am not. In my normal capacity as a writer, however, you may think me being irrational or illogical as I create these statements that are designed to reveal some underlying order within it all. Rationality is a slippery beast.

9.9 Lets stretch things a little further

Consider a particular state trajectory $\mathbf{x}(\cdot)$ underlying a statement that is the expression of a 'belief' held by its utterer. This progression of $\mathbf{x}$ can be renamed, say **Belief**(t). That being the case, then Equation (9.2) has the rather bizarre appearance of

$$\frac{d\mathbf{Belief}\,(t)}{dt} \tag{9.4}$$

A belief has changed over time. I have learned or I have reasoned or I have adapted in some way. Now consider the second derivative. Again, if we regard a particular trajectory as being part of a statement of belief, then we get the even more bizarre quantity

$$\frac{d^2\mathbf{Belief}\,(t)}{dt^2} \tag{9.5}$$

Beliefs are being forced, and are accelerating/changing. Where my beliefs were one thing, when asked, now they are another. Have I changed your beliefs about belief dynamics? I now know that unless I have a good dose of 1st and 2nd derivative I cannot function as a scientist. It is my job to account for novelty in a very formal way so that the resultant belief, some kind of communicable regularity in the natural world, when internalised into and guiding the state-trajectory of others, facilitates an increase in their scientific predictive power in respect of the natural world. This is the kind of nuance that arises in a complex dynamical system theoretic approach to learning. For our purposes here these concepts need be developed no further, but are recommended fodder for future ponderings.

9.10 P-consciousness within the state trajectory

The inclusion of P-consciousness in the state trajectory is simple. The state vector includes all brain regions, including those responsible for P-consciousness. To the extent that the state trajectory involves activation of the necessary brain regions, P-consciousness is generated. P-consciousness obviously exerts a controlling influence on us, possibly playing a role via equations (9.4) and (9.5).

9.11 Tokens, language and meaning

This discussion of statement production as a process of complex state dynamics has turned our understanding of language around somewhat. Consider the above discussion on 'belief', 'reasoning', 'intelligence' and 'rationality'. In each case it was possible, to some degree, to reconcile physical processes in our brains with the usual usage and intent of the words.

> An investigator who hoped to learn something about what scientists took the atomic theory to be asked a distinguished physicist and an eminent chemist whether a single atom of helium was or was not a molecule. Both answered without hesitation, but their answers were not the same. For the chemist the atom of helium was a molecule because it behaved like one with respect to the kinetic theory of gases. For the physicist, on the other hand, the helium atom was not a molecule because it displayed no molecular spectrum. Presumably both men were talking about the same particle, but they were viewing it through their own research training and practice.
>
> [Kuhn and Hacking, 2012, Page 51]

But I only say 'to some degree'. The reason for the 'some degree' qualification is that I have found real, observable brain processes that result in word production (as part of verbalised statements), yet the only time that word production has any meaning is when it is literally part of the state trajectory traversal. Pick any word you want and then place it in the state trajectory of 10 different minds, and then ask for a definition. You will get more than one answer (statement). It is only when the word is inhabiting and mediating the traversal of a particular brain's state trajectory that it actually acquires any worthwhile value.

In the case of scientists, we go to reasonable lengths (but do not always succeed, as Kuhn's example shows) to ensure that words are adequately calibrated. But the same words can elicit different behaviour between scientists, even for very formal concepts within the same discipline. The way science has handled this is to let the reality speak through *prediction*. If there is sufficient calibration of terms, then in a similarly trained engineer or scientist, the result will be that the externalised statement of regularity (say a 'law of nature') results in similar predictions. That is all science needs to do. Of course, the more technical specificity the better, especially if an intended recipient of a statement is located across a multi-disciplinary gulf. The basic neurological reality therefore only allows us enough latitude to think of words as *tokens* for the eliciting of targeted state-space trajectories in an appropriately trained recipient of the words. The phonemes or the syllables are symbols merely concatenated into tokens used for mental state trajectory control. All allusions to have externalised accurate meaning cannot be claimed by our understanding of how brains work. This also includes mathematical symbols, although this requires more specialised training.

This idea of statements being only tokens eliciting the fullest meaning in the brain of a receiver has direct connections to the discussions of wholes and parts in the next chapter (Chapter 10). What this position is actually saying is that the context of the statement is that of being part of a single, unified, inseparable 'whole' that is a system of perception, communication and the communication of perceptions (e.g. scientific observations). The extent to which such tokens ultimately influence the natural world is the extent to which the token interacts with the entire history (perceptions, knowledge and communications with others) of the receiver. While the measurable currency of any scientific community is the 'scientific statement' (a t_n that goes into set T), that statement has no meaning or separate life outside the context of it inhabiting a human during communication directed at learning or prediction or scientific critique/development. Kuhn clearly understood this. Another acquainted with this concept is David Bohm:

> Even more, it is generally only in communication that we deeply understand, that is, perceive the whole meaning of, what has been observed. So there is no point to considering any kind of separation of perception and communication. Perception and communication are one whole, in which analysis into potentially disjoint elements is not relevant.

> David Bohm [D. Bohm, 1974]

9.12 Building it

> Truth emerges more readily from error than from confusion.
>
> Francis Bacon (1561 – 1626)

The 'if you can't build it you don't understand it' theme has something to add here. I have some confidence in the complex-state-trajectory as a way of understanding cognition and the making of statements. For me, this is all obvious. The reason is that I have literally built so many 'state trajectory' machines I have lost count. All these machines operated in the real world. Industrial machines, process plants, whatever, their fundamental driving design is that of a state trajectory (called a state machine) identical in kind, if not in implementation basis, to the biological 'state machine' called the brain. In industrial control there is a high degree of parallelism, just like there is in our brains. There are inputs (sensory) and outputs (motor/actuator) just like in brains. It is easy for me to see how a brain is, at one level, merely a numinously sophisticated 'state machine'. I find this idea natural and compelling, simple and elegant. My long study and understanding of the biology of cognition from the whole organism to the atomic scale does nothing to tell me that I need consider brains to operate in any other way. Every measurement, every model of brain operation ever produced fits naturally within the conceptual framework of modern dynamics/complex systems theory. It's the easy problem.

Now for the hard problem.[b] What is not easy or obvious is why and how the biological 'state machine' that is a brain should also have a first-person perspective: P-consciousness. I cannot claim that anything I have ever built had this. It was not a design requirement. No existing understanding spoke to its presence or otherwise. As we have repeatedly found in these pages, science itself is completely unable to account for it. No law of nature ever produced by anyone predicts the existence of P-consciousness or prescribes some kind of non-optional role for it. For

---

[b] This encapsulation of the 'easy' and 'hard' problems in the science of consciousness is not new. David Chalmers has covered it some time back [Chalmers, 1995]. My use of the terms is essentially identical.

me, this lack is the show-stopper, and this book is about fixing it so that not only can complex system dynamics be exhibited by specific kinds of natural physics, but that same physics can somehow also be made to account for/use P-consciousness by revealing its role and level of necessity in different aspects of cognition.

I have already demonstrated empirically how science is configured *a priori* to be unable to tackle this 'hard' problem. My claim is that we literally define science not to be able to. So rather than make some kind of special 'Aha' discovery about the natural world as a normal act of science, I am going to rework science itself so that we can at least understand where a scientific account of P-consciousness might come from. Until we do this, a scientific account of P-consciousness will remain out of reach. We can now consider this event, in itself, as an act of state trajectory change that has already happened in me, and, I hope, may elicit such a change in you as you read this. Is there enough second-derivative in my words to make that happen?

9.13 Homework

What might be made of the following rather bizarre subsets of the state vector?

$$\frac{d\mathbf{Love}(t)}{dt} \tag{9.6}$$

and

$$\frac{d^2\mathbf{Love}(t)}{dt^2} \tag{9.7}$$

9.14 Summary

This chapter completed a depiction of science as being entirely a neurobiological phenomenon. This explains how centuries of scientific output can be posed, accepted and used and yet subsequently be overturned: it's all a product of the mind. To show this I delivered a complex dynamical systems treatment of humans as a form of 'dynamical agency'. It described our faculty for making 'statements' (novel and otherwise). One statement might be a serve in tennis. Another might be the vocalisation of "*f=ma*". The generic driver for the state trajectory, as revealed in behaviour, is the 'belief', considered to be merely brain configuration. Those brain configurations (some involved in P-consciousness generation) corresponding to abstract symbols are revealed merely as 'lookup tokens' for navigating brain state trajectories. The fact that such tokens might result in the trajectory following the progress of, say, the complicated grammar of mathematical symbols, is moot.

The key to it all is adaptive associative memory witnessed as a brain state trajectory. Viewed the right way, the navigating of state trajectories can be called reasoning. Changability and adaptability (the alteration of existing trajectories or development of new possible trajectories) was shown related to rationality and intelligence. I hope that you now have an appreciation for the way complex dynamical systems theory sees brain material, and how it presents the brain as a statement-making machine. In a book on science, the behaviour centred on making specialised explicit statements, the kind of complex dynamics I have depicted here is, to me, a defining characteristic of scientific behaviour. I do not imply that the scientist is a fantasy-maker. The statements made by a scientist are brutally held to account against the natural world. What I suggest is that to understand how we do science, we scientists need to stop projecting some kind of privileged, unique access to the natural world into our statements, and consider that our ability to make statements is telling us more about the natural world than the statements themselves. Applied to scientific 'belief', the process of depicting how statements come about completely disconnected us from all attribution of any meaning beyond that of being predictive of appearances (as detailed elsewhere here).

Chapter 10

Hierarchy, Emergence and Causality

The existing science framework (scientific behaviour) has been shown to be a neurological process incapable of providing explanation (causality) generally and an account of P-consciousness in particular. This chapter begins the process of finding out what science might look like if it could account for causality and explain P-consciousness.

> The whole is something over and above its parts, and not just the sum of them all.
>
> Aristotle (384 BC – 322 BC)
> Metaphysics (Book H)

Here I target an analysis of the science of parts and wholes for what it tells us about how science is/has been conducted and how it may be conducted. The literature on parts and wholes is vast, old and confused. It is full of attempts to grapple with words like 'irreducibility', 'unpredictability', 'conceptual novelty', 'holism' and a favourite here, 'unexplainability' [Bedau and Humphreys, 2008a]. As with consciousness, discussion started largely outside science and has become more science-centric in recent decades. Again, this is consistent with a Kuhnian route to a shift of some kind. Where it impacts science, the modern treatment of parts and wholes travels under the umbrella of the words hierarchy and emergence. It is most visible in the science of complex systems. We have seen how complex the brain is. The link is rather obvious. I will focus mainly on the idea of hierarchy because emergent properties naturally 'emerge' from a treatment of hierarchy. For my purposes here I need not attempt to grapple with anything else. I do not seek a solution to emergence. What I deliver is an account of how science, as presently conducted, is predictably confused about emergence.

It is now clear I am proposing a new science framework. Its practical form and execution is developed in Chapter 11 and is called 'Dual Aspect Science' (DAS). To establish the nature of the change to science, here I directly confront the explanation of the 'whole' that is the scientist. By 'explanation' I mean formal causality; that which is missing from science at the moment. The hierarchy issue is a natural fit to this task and I will show you how. Within the upgraded framework is a potential account scientific observation. It is naturally appreciated from the perspective of a treatment of hierarchy. At the moment, science is correctly classed as 'single-aspect'. How can I possibly propose a second aspect when the reader is unable to mentally conceive of what the second aspect actually is? Considerations of the scientific treatment of a hierarchy of accreted parts (that is, a whole) can make plain the reality of single-aspect science in a way in which the new, second, aspect is abundantly clear.

Turned around, what this means is that the DAS framework is a natural solution to the problem of parts and wholes, and therefore of what has been called emergence. It does this by eliminating the problem, not by providing a specialised technical solution. The problem of emergence is, under the DAS framework, a non-problem related to the perspective and traditional 'single-aspect' habits of scientists. Please note that only the bare minimum is covered here – sufficient to address the main issue of the targeted problem, a scientific explanation of the scientific observer. It is left to others to develop the DAS framework's perspective on the more common problems posed in the area.

The second 'aspect' in the DAS framework is a significant conceptual leap. Consideration of parts and wholes can be used as a vehicle to encounter both (i) the reality of the scientist's lot under the upcoming dual aspect science and (ii) the universe as it must be to the dual aspect scientist. For that reason, this chapter is probably the most important chapter in the book. It forms the basis for a new way of conceptualising the natural world. As you read this chapter, if you find your mental grip hitting a wall, and that wall telling you that the discussion must be invalid, the Kuhnian wisdom is that I find the world of that concern equally impenetrable. You and I are in different worlds. A large part of this book is about spanning that gulf, and I suggest that you try to find a place where my view makes sense in spite of your concerns. That will be a place that the new and the old can meet productively. That place will ensure adequate critique is applied.

For the scientist, analysis of natural organisational hierarchies creates a perspective that is intrinsically multidisciplinary. Multidisciplinarity itself is one of the major hurdles the scientist faces. No single science subdiscipline rules the roost. Consider the natural hierarchy of our descriptions etc...$\rightarrow$ atoms $\rightarrow$ molecules $\rightarrow$ cells $\rightarrow$ .... population $\rightarrow$ ecology $\rightarrow$ ... $\rightarrow$ etc. This clearly connects physics, chemistry, cell biology, all the way through psychology to social sciences, to ecology and upwards to cosmological concerns, and so forth, in a single narrative. Because of the cross-disciplinary nature of the study, it has no obvious scholarly home, no obvious distinct community of practitioners, and the literature reveals that its contributors tend to be spread around the precincts of special interest groups grappling with a chosen part of the hierarchy. A Kuhnian perspective on this tells us that, at this point in time, the science is in a pre-paradigmatic state. For the purposes of argument and critique, I claim here that the pre-paradigmatic state of hierarchy and emergence is actually a symptom of the pre-paradigmatic state of the science of scientific behaviour. To see this you'll have to ride through this chapter.

Before I start I have to reinforce the reality of single-aspect science as revealed in previous chapters. Most scientists will find it difficult to encounter the fact that single-aspect science, of the kind done for hundreds of years and that is the bread and butter activity for us all, never touches causality – that which necessitates that the natural world unfolds the way it does. We may assemble 'sufficient conditions' to bring about some desired natural outcome, say X, and be 100% confident that we will succeed. But being able to identify and establish 'sufficient conditions' is only to be able to assemble the *appearances* of those sufficient conditions. There is no actual explanation of why X happens. We never do that. Knowledge of the sequential orders/concomitance of observable phenomena does not carry with it any explanation of what it is about the underlying natural world that makes it so. If there is one thing that the reader must internalise, this is it. If you can't see this, then dual aspect science will make no sense.

There is an interesting link between parts, wholes, a single-aspect scientific account and their relationship to engineering/technology. Classical single aspect science reduces the natural world to parts. A scientist chooses a whole, and then a part is declared as the target of study. Conversely, a whole might be studied by literally ripping it into parts. This is called reduction. I call it single-aspect. The process of reduction can be abstracted a bit by realising that what a single-aspect scientist does is observe the natural world as it appears (a whole), and then work backwards to the initial conditions (the natural regularity that is the behaviour of its apparent parts). In the process, there is no connection with the underlying causality that makes collections of observed parts act as the observed whole. All that is discovered is *that* it happens. It's all 'what' and no 'why'. All description and no explanation.

Engineers, on the other hand, work in an inverse fashion. For engineers there is some kind of desired state of affairs (such as an automobile's ability to transport). It is the engineer's job to establish the initial conditions (design/build a car). Based on an understanding of the sequential orders or concomitance of phenomena, the engineer assembles parts, and the desired result becomes the 'whole' which, hopefully, is on a trajectory provided by the initialised parts. Single-aspect science, and single-aspect engineering are therefore a complementary pair of behaviours, and all of it operates without actually knowing why the natural world makes it happen. The rules of concomitance and sequence that we call 'laws of nature', and that correlate observed phenomena, facilitate this process. We don't have to know why. We only need to be predictive of appearances. We are merely guided by 'apparent causality' evident in observed phenomena: our experience of the natural world.

Your challenge, then, is to encounter the possibility of a science that actually deals with causality. This is the second aspect. Isn't it obvious now how single-aspect science can never predict or explain an ability to scientifically observe? Isn't it obvious now that science has never actually 'explained' anything, let alone a scientific observer? Remember: recognising that single-aspect science never 'explains' (in the technical sense of causality) does not entail that scientists cannot account for causality. That single aspect science does not/cannot predict or explain an ability to scientifically observe does not entail that scientists are fundamentally prevented from ever doing it.

10.1 Aside: Causality (causation) vs. critical dependency

All else being equal, a critical dependency of Y on X is revealed if it is necessary for X to occur before Y can be observed. Some writers use the term 'constant conjunction'. Some use the term concomitance. Others might say that 'Y supervenes on X'. A critical dependency reveals the outward signs of such a causal relationship, but not what necessitates that it is so. This basic position was established hundreds of years ago by David Hume[Hume and Steinberg, 1993].

10.2 The second aspect: A fundamental challenge

I now seek to draw a picture of the world in which the second (new) aspect of scientific description is made plain. I draw your attention to scientific knowledge viewed vertically across all the disciplines. In Figure 10.1(b) we see a hierarchical depiction of the layered organisation of the *observed* natural world (whole) that includes the scientist/observer (part). Figure 10.1(a) shows the vertical assembly of scientists that target each of the layers (parts). The hierarchy itself has been chosen to include that of the scientist's brain – that special 'whole' that makes all of this possible.

In the hierarchy of the natural world, science is conducted within Figure 10.1(b) at layer (M+4) by a population of humans called scientists. That does not entail that science can only occur in groups. An individual scientist can do science (Figure 10.1(b) layer (M+3)). Group behaviour only makes it better, more effective science. The fundamental properties of human cognition, that make the behaviour possible, exist at Figure 10.1(b) layer (M+2).

The 'organisms' that are directly charged with explaining Figure 10.1(b) layer (M+2), scientists, are neuroscientists and are located at Figure 10.1(a) B-B′. Currently that science is called the science of consciousness and, as previously discussed, implicitly (methodologically) denies that it is there to explain scientific observation and scientists. The rest of the Figure 10.1(a) collection of scientists (those not doing the science of consciousness) proceed by producing scientific statements that are in-principle prevented from predicting or explaining an ability to observe. Rather, it predicts how the universe will appear to an assumed observer. This proceeds smoothly, without any scientist involved in the process being aware of the fundamental limitations inherent to the circumstances of it.

Now we have our first glimpse of parts, wholes and hierarchy. What we don't yet have is a depiction of or connection to underlying causality. Consider the cell biologist studying Figure 10.1(b) layer (M) – the cell. A cell is entirely composed of 'parts' called molecules (Layer M-1), and becomes part of an assembly (a whole) of cells called tissue (Layer

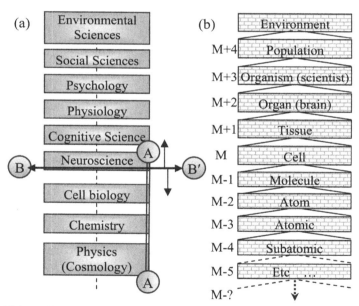

Figure 10.1.   The observed hierarchy of the natural world, (b), and the hierarchy of scientists that create it, (a). The matter hierarchy of the brain of the scientist is depicted.

M+1). There are many parts and wholes, and there is no hard and fast rule for their isolation and naming. The main criterion is that a scientist can observe some kind of regularity. It is the regularity that interests us, not the part/whole context. It is helpful to have a clear delineation and a name, but only for the purposes of accurate communication.

Now change perspective. Consider our position as scientists-of-the-hierarchy Figure 10.1(b), using the faculties of observation acquired from being of-the-hierarchy to create statements-of-the-hierarchy Figure 10.1(b). How does this ability to scientifically observe arise, as a whole, within the hierarchy? One way to see this is to look at each layer. Technically the layered hierarchy up to Figure 10.1(b) layer (M+3) is called a 'containment hierarchy'. That is so because each layer is entirely and only comprised of members of the lower layer(s). A tiny sample of the historical depiction of hierarchy and emergence is a good way of encountering this issue. While the mystery of part/whole emergence is very old, science itself didn't really start to grapple with it until the scientific disciplines themselves were largely defined and stable, which really began to be the case in the second half of the nineteenth century:

> ...We do not suppose that when what is called the physical motions of molecules are grouped into what is called chemical actions, and surprisingly novel phenomena emerge, there has been anything essentially superadded to the primitive molecules and their forces...The chemical phenomenon is new, the vital phenomenon is new; but the novelty is one of special grouping of the old material and the old energy.
>
> [Lewes, 1879, Page 189]

> All organised bodies are composed of parts, similar to those composing inorganic nature, and which have even themselves existed in an inorganic state; but the phenomena of life, which result from the juxtaposition of those parts in a certain manner, bear no analogy to any of the effects which would be produced by the action of the component substances considered as mere physical agents. To whatever degree we might imagine our knowledge of the properties of the several ingredients of a living body to be extended and perfected, it is certain that no mere summing up of the separate actions of those elements will ever amount to the action of the living body itself.
>
> [Mill, 1930, Bk.II, Ch.6, §1]

Lewes and Mill tell us that if you take away the cells in Figure 10.1(b) layer (M), say, then everything is gone. There is nothing else. Take away the cells and you have no organ. No organs = no organism. Going deeper in the hierarchy, if you take away the molecules you have no cells. No cells = no organs. The result is the same. This view is a bit simplistic, because in any one instance of a member of a layer, exactly what is included by the word, say 'cell', is not obvious or simple to define. But we all have an intuitive grip on what it means to delete a 'cell' or delete 'all cells'. A cell can be thought of as a bunch of molecules embedded in space. There is nothing else to delete. When you're done, only space is left (notionally). However, when you deleted the molecules, you not only eliminated their mass, charge and the fields they express, you also deleted a pattern – the unique organisation of molecules in space. To restore the molecules requires more than just identification of molecules and their quantities – it requires detailed locations and interrelationships. When you delete any layer, in addition to the obvious material/field systems, an enormous amount of information (Lewes called it the 'special grouping') is also deleted. 'Information', however, is intrinsic; not something you can easily point to. In terms of the 'magic' or 'elan vital' that makes the whole more than the sum of the parts, clearly it must be related to the information content, but not in any way that is simple to articulate at this stage.

A final way of characterising the containment hierarchy is by illustration of another kind of hierarchy. The classic exemplar is the military hierarchy. In the military hierarchy the commanding general is not literally composed of the component troops. Yet the behaviour of the troops is determined by the commanding general. In a sense the collective behaviour of the troops is not a property of the collection of troops. The imperatives of the group are, in a sense, outsourced or exogenous.

At this level, containment sounds simple. But an extended hierarchical view reveals a profound implication. Go back to Figure 10.1(b) and consider the deeper layers (M-3), (M-4), (M-5), (M-6), (M-7),. ... (M-10,000),... (M-?). How many layers are there? We know that if you delete (M-*anything*), then every layer above, all the way up, disappears. Physics tells us, so far, about the layer (M-4), subatomic. But the fact that humans have been able to successfully interact with that layer and thereby identify it, in no way entitles us to conclude, all of a sudden, that we have reached an absolute limit. Conversely, the fact that we have been able to interact with the structure, using the structure, in no way allows us to conclude that the natural world is made of isolated, disconnected processes of any kind. Indeed, the 'deletion' thought experiment, with its indefinite propagation of vertical disappearances, tells us this cannot be so. One solution to this conundrum is shown in Figure 10.2. All layers derive from a single large collection of structural primitives that self-organise into a single unified whole, from which space is expressed (M-J) that serves to contain other layers of organisation above, that apparently act like atoms, molecules and so forth. We observers at location (M+3) see it that way. The entire thing contains itself. The entire thing is a single unity. The entire thing is composed of a structural primitive of some kind (only one is needed in the simplest case) and every single one of these structural primitives is connected to others, and there are no exceptions. This is the obvious logical endpoint if you extend the self-containment idea while requiring no isolated parts, yet have a unified system that has an ultimate smallest-component.

Time out! Step back and take a look at where we are now, as a result of a simple discussion that takes a view most scientist's don't take: (i) cross-disciplinary and (ii) as an observer embedded inside it. From this perspective the universe itself has a totally different character, and that character, as seen from the confines of any individual science discipline, looks exactly as it did before. *You have now encountered the 'second-aspect' for the first time.* This is the proposed major shift in approach. In the second aspect you are required to stop describing the natural world in the way science traditionally describes it (appearances, apparent causality). Instead, the natural world is to be described as a collection of structural primitives operating as a single, gigantic, seamless process throughout; a single *indivisible* whole. This can be true without compromising any existing (single-aspect) scientific view. All that has to happen is to construct a way of exploring the vertically-structured unity. This is what is detailed in the dual-aspect science chapter. It's not particularly hard.

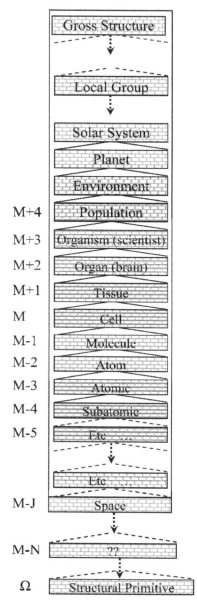

Figure 10.2.   The probable final state of the hierarchy, in which the ability for the entire structure to contain itself is preserved.

The science of wholes runs counter to our natural inclinations as observers weaned on the science of parts, but this can happen without invalidating the single-aspect (science of parts) approach. This view of nature (as a single indivisible whole) is far from new. People such as David Bohm did this years ago [David Bohm, 1981]. What is new, however, is the simple recognition that the scientific statements that come from existing single-aspect science, and the scientific statements that result from the new 2nd-aspect are not the same kind of statements. They are not merely a continuation of the kinds of statements produced at each layer by observation/interaction. Nor are they incompatible with each other. Nor does one have higher status than the other. Two different kinds of statements about the same natural world.

Viewed from the second-aspect, we can see that words like 'hierarchy' arise as a result of description of the underlying natural unity from within it. If you start from this position you can see immediately that, however we break down layers for descriptive purposes, the organisation of the natural world can only claim to have an *apparently* hierarchical organisation. That claim does not entitle the claimant to say that the natural world is literally made of the descriptions thus constructed. In other words, the natural world is not made of anything in the hierarchical description. It's not made of atoms or space or photons or electromagnetic fields or anything else of that kind. Rather, it is made of something that has that scientifically documented appearance when you are made of the same 'stuff'.

The difference between these two things is the gulf that we are exploring here. The challenge for the reader is to be able to see that difference, and be advised that the underlying stuff is just as observed as anything else, just not directly. The mistake of the past was to assume that just because the underlying natural world is not directly 'scientifically observable', that it is has no 'scientific evidence'. To resolve this conundrum involves an account of the scientific observer, that must, by definition, be made of the same '*stuff*', and therefore be able to observe by virtue of being made of that '*stuff*' inside a massive unity of it.

Paradigms are not corrigible by normal science at all. Instead, as we have already seen, normal science only leads to the recognition of anomalies and to crises. And these are terminated, not by deliberation and interpretation, but by a relatively sudden and unstructured event like the gestalt switch. Scientists then often speak of the "scales falling from the eyes" or of the "lightning flash" that "inundates" a previously obscure puzzle, enabling its components to be seen in a new way that for the first time permits its solution.

[Kuhn and Hacking, 2012, Page 122]

10.3 Causality, the 'Ghost in the Machine'

The cross-disciplinary view reveals that we can accept that there is an underlying reality, and that it is a single indivisible unity, immediately clarifies causality. To explore causal relations now has nothing to do with correlating observed phenomena, but involves tracing linked underlying structural primitives. One of these tracings of structural primitives will reveal the underlying workings of the scientific observer. To see how this works requires some more conceptual tools. To start with I will adapt a very recent depiction of situated cognition that maps easily into a treatment of hierarchy. Randall Beer, in 2004/1995 [Beer, 1995, 2004], produced the diagrams shown in Figure 10.3(a). There is a primitive correspondence between it and the Figure 10.1 hierarchy of the scientist. What is missing from diagrams Figure 10.3(a), however, is the observing scientist (in this case, Randall Beer, staring down at the diagrams, unconcerned that he was inside what he was drawing).

Figure 10.3(a) has some of the hierarchy in it, but it depicts it in a way that is in-principle impossible under the system of unified hierarchy proposed by Figure 10.2: the disconnectedness of the layers. Diagrammatically it assists to delineate the layers, but it has no physical justification. I have taken the opportunity to develop Beer's Figure 10.3(a) diagrams a little. I put the observer *Scientist(·)* inside the environment with that which is observed, called *Studied(·)*. The notional boundary is now seamless with the containing environment, and the boundary is recognised as a source of measurement (sensory/affect) and control (motor/effect).

(a)

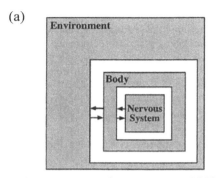 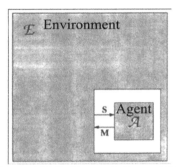

[Beer, 2004, Figure 5. Page 320]   [Beer, 1995, Figure 2. Page 182]

(b)

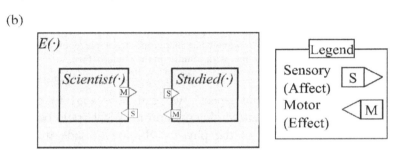

Figure 10.3.   The delineated boundaries of 'agency'. (a) The historical way of depicting natural hierarchy. (b) Revision to account for the reality of hierarchy studies by scientists.

Next, I can recognise the term 'ghost in the machine', made popular by Arthur Koestler [Koestler, 1967], to more formally illustrate the actual hierarchy. The 'ghost in the machine' has already been seen in the above Lewes/Mill quotes, which allude to 'vital forces'. This is evidence of a physics/biology rift directly attributable to the mystery of the part/whole relationship. The belief in the mysterious 'elan-vital' is called vitalism. Even now, well over 100 years since Lewes/Mill, the mystery of how physics parts make a living biological whole remains alive and well. Consider the following 1999 apologetically quasi-vitalist musing within a membrane biology journal, under the heading 'vitalism and complexity':

> The structure and underlying physics are not always enough to understand biological systems of complexity, because the complexity itself adds qualitatively

new behaviours underlying pieces of the system. Whilst these behaviours are certainly compatible with the underlying physical laws of the pieces, and in that sense implicit in them, they cannot be uniquely predicted from those underlying laws without a detailed understanding of the hierarchy of structure. In many fewer words: a machine does much more than its parts do separately because its parts are designed are designed to work together to perform a function..... in this quite limited sense, explanations are needed for biological systems that lie outside the laws of physics, as they are usually presented. The explanation must include both physics and structure, but it cannot consist only of the structure or only of the physics, at least in my view.

In this quite limited sense then, vitalism is an appropriate part of biology. Physical laws undoubtedly govern the behavior of these complex systems, as well as governing the behaviour of their elements, but, taken as a whole, biological systems, and organisms, often show behaviours behaviors that reflect the hierarchies of structures more than the properties of the elements of those structures. Those behaviours might be called 'vital', organic to the complexity of the structure, not obvious in the underlying physical laws... ... Thus the word 'vitalism', which I used above us somewhat inappropriate, a piece of artistic license, that I hope may be granted me, with a smile on the reader's face.

[Eisenberg, 1999]

By recognising the 'second-aspect', we can now start to get an inkling of the origins of the part/whole conundrum. What could be more indicative of the recognition of the physics⇔biology divide than the name of Arthur Koestler's well known 1967 book 'The Ghost in the Machine'? This is our connection to the history from the 1960s to 1980s. It is in this era that some of the more useful and usable hierarchy concepts were prototyped. Koestler's book established a dialogue that had a significant impact throughout the era of Thomas Kuhn's *Structure*. Koestler (in his 'The Ghost in the Machine') finally realised that simplistic approaches were inadequate:

> The first universal characteristic of hierarchies is the relativity, and indeed ambiguity, of terms 'part' and 'whole' when applied to any of the subassemblies...
> ...A 'part', as we generally use the word, means something fragmentary and incomplete, which by itself would have no legitimate existence. On the other hand, a 'whole' is considered as something complete in itself which needs no further explanation. But 'wholes' and 'parts' in this absolute sense just do not exist anywhere, either in the domain of living organisms or of social organisation. What we find are intermediary structures on a series of levels in an ascending order of complexity; sub-wholes which display, according to the way you look at them, some of the characteristics commonly attributed to wholes and some of the characteristics commonly attributed to parts...

[Koestler, 1967, Page 65]

Arthur Koestler approached hierarchy with the 'holon' in a treatment of hierarchy that would now be called 'nested' because the hierarchy arises through physical containment. It still took some time for the social and non-social hierarchy to be properly distinguished and related to containment. The development occurred in ecology, where the nested (containment) and non-nested distinction was added to the Koestler 'holon' conceptual framework by Allen and Starr, who were grappling with the complexities of large systems. They used Koestler's nomenclature:

A nested hierarchy is one where the holon at the apex of the hierarchy contains and is composed of all lower holons. The apical holon consists of the sum of the substance and the interactions of all its daughter holons and is, in that sense, derivable from them. Individuals are nested within populations, organs within organisms, tissues within organs, and tissues are composed of cells. ... .Nested hierarchies meet all the criteria of the more general hierarchical condition. The higher holons in the system are associated with slower time constants of behaviour(i.e. longer cycling times and relaxation times), and, if manifested in space, they occupy a larger volume. This is common to all hierarchical construction. ...

[Allen and Starr, 1982, Page 38]

Note that very little modern usage of the 'holon' terminology can be found in the sciences. The only significant live example in recent literature is in the applied sciences related to manufacturing where product has been treated as an emergent property of a production system. These systems are called 'holonic control/production systems'. The holon concept is a natural fit where large distributed systems cooperate to a single end. The modern distributed, networked/hierarchical production environment is good example of living hierarchy (holarchy) theory applied to artificial systems. Koestler's contribution to understanding nesting was perhaps more significant because he recognised the intrinsically dualistic descriptive milieu that is implicitly created with hierarchy:

The members of a hierarchy, like the Roman god Janus, all have two faces looking in opposite directions: the face turned towards the subordinate levels is that of a self-contained whole: the face turned upward towards the apex, that of a dependent part. One is the face of the master, the other the face of the servant. This 'Janus effect' is a fundamental characteristic of sub-wholes in all types of hierarchies.

[Koestler, 1967, Page 65]

This fact was not lost on Allen and Starr:

> Koestler uses the image of a doorway between parts of the structure and the rest of the universe. The entity has a duality in that it looks inward at the parts and outward at an integration of its environment; it is at once a whole and a part. At every level in the hierarchy there are these entities, and they have this dual structure. As in taxonomy, where each level in the hierarchical classification is called a "taxon", Koestler calls his two faced entities "holons".
>
> [Allen and Starr, 1982, Page 9]

I have used Koestler's depictions of the levels of description, their containership state and the 'Janus view' in Figure 10.4. The hierarchy has again been defined with brain material in mind and with the human observer embedded within the hierarchy as a member, trying to describe the hierarchy using observational faculties inherited as a result of that very circumstance. It is this fundamental fact that is taken from Koestler into the rest of this treatment of hierarchy. What is novel is that I have

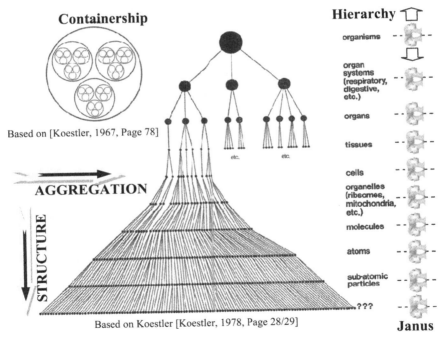

Figure 10.4.   Koestler's 'Janus-faced holarchy' and its axes of expression.

named two axes. The structure axis indicates the direction towards smaller spatial scales and more primitive organisational levels. Organisational accretion along the structure axis gives rise to layers. The aggregation axis represents the simple agglomeration of members of a layer. The 'horizontal' axis expands a layer by accumulation, at a single spatial scale.

The 'Janus' idea can be thought of as the 'vertical' version of the 'horizontal' S/M boundary of Figure 10.3. This is the beginning of a more sophisticated view of causality that is intrinsic to the dual-aspect model. Vertical causality acts along the structural axis. Horizontal causality operates along the aggregation axis. With this view it is immediately obvious that what science has been doing all along is describing short-hop *horizontal* (aggregation) relations (such as electrons meeting a nucleus or an atom meeting an atom) consistent with the reduction of a localised whole into parts. The next obvious observation is that layers are themselves simply excursions along the aggregation axis that then acquire their own stability, and a net affinity for other entities at the same spatial scale. For example, a bunch of atoms that form a molecule. They all exist at a certain spatial scale, and they all operate as a single entity at that spatial scale. As a whole, they interact with other molecules to make, say, a cell. And so forth up the hierarchy. This is the 'alternate horizontal/vertical causality' that couples members of a layer together to form a whole novel structure.

To see this layered aggregation more explicitly, consider another representation of the Figure 10.1 hierarchy shown in Figure 10.5(a). Because each layer is an aggregation of the members of the lower layer, the number of members of the higher layer is reduced. This one of a number of general properties of hierarchies formally captured in Feibleman's 1954 'Laws of the levels' [Feibleman, 1954]. We need dwell on these no further here.

In Figure 10.5 I have generalised the 'sensory'/'motor' boundary interaction to accentuate the trail of causality from the deeper layers. Yet another representation of the same basic idea, but generalised, is shown in Figure 10.5(b). In this instance, the natural causal affinity (S/M) for three layer N entities combine to form a new layer, N+1, which has properties (S/M) of its own. The reader is encouraged to internalise the idea that no matter how complicated the affinity at each layer, the layer includes *all the layers there are, all the way to the bottom of the Figure 10.1 hierarchy* (shown as layer Ω in Figure 10.2 or Figure 10.5(a)).

The implication is that there is an unbroken structure-axis linkage from all members of the hierarchy to all others, and this is via the connections between the layers all the way down to the bottom. In this way we can now start to see how the 'Janus' linkage is real, and physical.

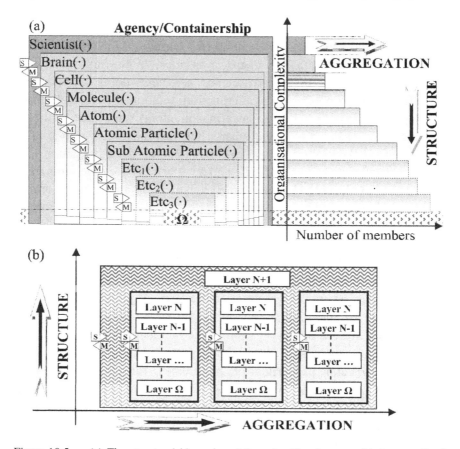

Figure 10.5.　(a) The structural hierarchy of the scientific observer. (b) A generalised aggregate of three layer N entities to form a new layer N+1. Once formed, layer N+1 has an affinity for further layering and interaction that is indicated by its own S/M. Note that layer Ω is shown differently because potentially connects all layered systems to each other.

It's just not a physical linkage we are used to considering. Every layer in the hierarchy is literally composed of successive layers of *organisation* of the single layer of a hyper-aggregated unity that is the very bottom layer of the Figure 10.2 hierarchy. This means that causality is actually much more complex than the simple horizontal (layer ⇔ layer) interaction characterised by single-aspect science. Wherever two entities X and Y meet and interact in the fashion that we are used to, it causes a horizontal X⇔Y connection that closes a loop from X all the way down to layer Ω, across, and then up Y. All material entities have this potential shared linkage via layer Ω. Layering of this general kind, and layer Ω in particular, changes our view of the 1PP. By 'being' a collection of any layer, we automatically inherit a potential linkage to all other layers via layer Ω. We must 'be' all of it, all the way down to layer Ω. We cannot avoid it. Clearly this fundamentally alters our relationship with all other members of the hierarchy, especially those entities with which we can interact and 'close the causal loop'. This is the basic physical reality upon which one can build an account of the 1PP. That account, whatever it is, becomes completely consistent with, while being separate from, all single-aspect scientific knowledge of 'apparent' causality.

The final depiction of causality, and the nature of linkages between entities (the layered material and its field systems), is shown in Figure 10.6. It depicts the layering of two cognitive agents with a brain that share an environment. Consider it a more complete version of Figure 10.3(b). The subtlety that emerges in Figure 10.6 is at the layer called 'space'. Because space is a container of all the layered material of the higher layers, and is itself made of the identical collection of the layers below (all the way to layer Ω), every form of layered agency embedded in space/part of space, is connected, through the space layer, to every other agency. All of it is ultimately connected at layer Ω.

Probably, at this stage of proceedings, to many scientist and layperson readers, this may look like new age everything-is-connected space-cadet mumbo-jumbo. Well all I can say is: *patience*. The current state of science – the single-aspect – remains utterly unaffected by any of it. No science to date can deny the possibility of the second aspect, and the entire thing is designed to account for something that single-aspect science can never do: predict and explain the scientific observer (= you,

the reader). The clinching proof that this approach is valid, is us –
scientists. We are the proof that this new way of thinking is valid. And
we get to prove it empirically. This will be covered in the chapter on dual
aspect science and the chapter on testing for consciousness. It is upon the
consciousness of scientists that this entire proposition is critically
dependent. So I am asking those scientists for which this kind of
presentation seems rather farfetched, to bear with me for the rest of the
book. I had to go through the exact same process of incredulity. Just
remember that the two aspects in the proposed 'dual-aspect' science
framework, are tied at the hip in an explanation of scientists and
scientific observation (P-consciousness).

To do the science of the second-aspect is to use techniques covered in
Chapter 11. Figure 10.6 reveals the special status of a particular organ
(the brain), which we know uses properties of layer M (cell) to construct
P-consciousness. To then use that faculty to scientifically observe and
describe the appearances (such as cells, molecules, space and so forth),
leads to the single-aspect science we have been doing all along.

Single-aspect science description is, then, a result of a *horizontal view*
at a layer or between layers. The causal relations we normally depict
(such as $f = ma$) account for the observed horizontal interactions that we
see all around us. This is only 'apparent' causality. In reality, all the
causality originates in the vertical direction as merely organisations of
layer Ω structural primitives. There is therefore a meshwork of
effectively circular causality all the way from the peak of the
organisation to the layer Ω base. To see how this works, remember that a
cell is merely a collection of molecules, and molecules are a collection of
atoms, and so forth, all the way to the bottom. There are actually no
atoms, no molecules. The entire thing is a collection of layer Ω structural
primitives connected in a seamless chain throughout the entire vertical
layer system. This is the simple reality revealed by the thought
experiment of the start of this chapter.

Notice that, by virtue of the special containment status of the 'space' layer, everything expressed by space, up through all the higher layers, has a special relationship with space. It is rather obvious that this is the first place to look if trying to account for what the 'cell' layer is doing in brains in order that scientists are P-conscious. Remember: we scientists get to 'be' all the layers, top to bottom. All we are ever 'being', however, is a collection of layer Ω structural primitives.

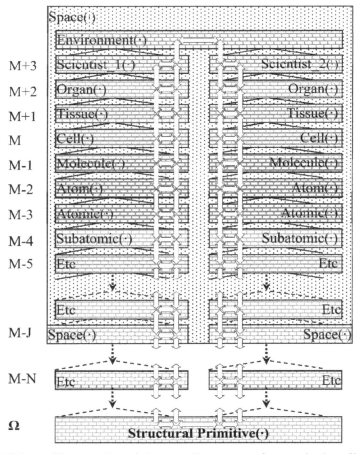

Figure 10.6. The causality of the overall structure of two scientists. Horizontal linkages are between members of the same layer. The vertical linkage is a direct connection to members of Ω layer whose organisation form the layer. Causal connections loop down from one hierarchy to Ω and then up another hierarchy and across.

10.4 Dynamic hierarchies

What has not been shown in this discussion of hierarchy is that the hierarchical structures of the kind illustrated above are intrinsically and universally dynamic. This means that all organisational layers are best thought of as persistent structure expressed by componentry whose fundamental nature is one of change. For example consider the layer called 'population'. Population is a valid structure comprised of the collection of all humans, each of which exists for a finite time. The structure 'humanity' persists when its human components do not. In reality there is no layer in any of the depictions so far that is not dynamic to some extent over an identifiable timescale.

(a)

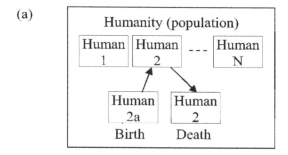

(b)   The nested dynamic hierarchy

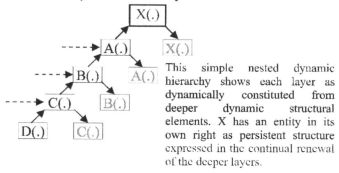

This simple nested dynamic hierarchy shows each layer as dynamically constituted from deeper dynamic structural elements. X has an entity in its own right as persistent structure expressed in the continual renewal of the deeper layers.

Figure 10.7.   Dynamic hierarchy. In (a) we see how humanity is a temporary persistent collection of its deeper layers (humans). The population can persist while the individual members of the population do not. In (b) we see how an arbitrary group of deeper processes cooperate to create a persistent structure $X(\cdot)$.

The same thing can be said to apply for atoms, which all have a finite probability of splitting into smaller atoms. Their stability is merely relative. Atomic particles such as a proton have a finite (but very large) half-life. A chemical equilibrium such as an electrolyte solution is an example of a dynamic hierarchy at the molecular level. This dynamical stability has also been observed at the level of very large molecules in a brain matter context [McCrone, 2004]. Cells in the human body are routinely replaced while the organs they are part of continue to function. There are no exceptions to the dynamic nature of naturally occurring hierarchies. All appearances of stability of an entity are no indication of the actual stability of the entity's components. There can only ever be claimed to be sufficiently persistent structure to allow the observed natural behaviour. Probably the most often quoted and classic example of the dynamic hierarchy in nature is the candle. The candle flame remains as persistent structure made of moment to moment change, exactly as shown in Figure 10.7.

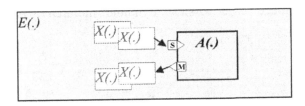

Figure 10.8.   Dissipative dynamic hierarchy.

10.5 Dissapative and lossless dynamic hierarchies

In considering the full expression of the interaction between a dynamic hierarchy (say, called an agent) and its environment, one must account for everything impacting or entering the bounds of the agent. This is shown in Figure 10.8. The 'sensory' (S) ⇔ 'motor' (M) boundary for agency $A(\cdot)$ must, in general, account for absolutely everything that is transacted with the enviroment $E(\cdot)$. In other words the behaviour of $A(\cdot)$ is not merely machination of a physical boundary. A complete inventory of S and M would include thermal energy (energy of motion) in/out, electromagnetic fields/energy in/out, matter in/out and gravitational energy in/out and whatever is involved in the occupation of space.

In the case of a human the components of S could be nutrients, water and oxygen along with biological insults such as bacteria and electromagnetic phenomena such as heat and incident radiation of all kinds. M would include all excreta, exfoliated cellular material and respiratory byproducts along with electromagnetic phenomena such as heat, reflected light and sound.

If the environment E(·) has to continually deliver energy (via S), in whatever form, into the agent in order that the agent structure continues to operate: this is a dissipative structure. A system that necessarily takes on and dissipates energy as it interacts with its environment is a dissipative system. 'Dissipative' does not mean any collapse into disorder. Dissipative systems maintain identity because they are open to flows of energy, matter, or information from their environments [Prigogine and Stengers, 1985, 1997]. A photon is a lossless dynamic hierarchy. Space is a lossless dynamic hierarchy. A human is a

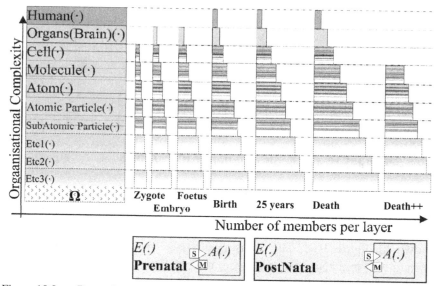

Figure 10.9.   Dynamic agency. The autopoietic dynamic agency that is human physiology. It reveals changes in 'agency', as alterations to the number of layers, over the life cycle of the human.

dissipative dynamic hierarchy. A dissipative agent A(·) can operate far from thermodynamic equilibrium with an environment E(·) with which it exchanges energy and matter under the entropy drives of the second law of thermodynamics. A dissipative system is characterised by the spontaneous appearance of symmetry breaking (anisotropy) and the formation of complex, sometimes chaotic, structures where interacting particles exhibit long range correlations and organisation that would be impossible without continual energy supply. This is the far-from equilibrium circumstance of the brain that results in the complex dynamics discussed elsewhere in this book.

The word 'dynamic' normally implies change and has so far been used somewhat paradoxically to express lack of change (stability) in the form of the persistence of the dynamic hierarchy described above. What is perceived as the hierarchy is merely and only that persistence. The term 'dynamic' in this context *does not mean that the hierarchy is dynamic* (changeable), but that the hierarchy itself is created and maintained dynamically i.e. with componentry that is refreshed continually or better, literally *is* change. This distinction is crucial if we are to handle the full description of brain material.

10.6 Hierarchy change dynamics (dynamical agency)

Where we directly consider the additional dimension of change, we can look at actual alterations to the hierarchy. Note that in Figure 10.1 the layers are static in the sense that they have a defined existence even if instances of the membership of the layers are dynamically created by lower layers in the hierarchy. The structure is dynamically refreshed but the agency remains constant. The layers remain. That being the case, how do we address autopoietic (self-constructed) and allopoietic (other-constructed/artificial) hierarchies? This kind of change means adding, subtracting or otherwise altering layers so that agency is altered. This is the inevitable outcome in the dissipative agent operating under the second law of thermodynamics that develops from a simpler state to a more complex state (such as embryonic development) or through physical manufacture. The changing of the number and composition of layers can be called 'dynamical agency', in contrast to the 'dynamical hierarchy' dicussed above. The distinction between dynamical agencies and dynamical hierarchies is a slight development of work by [Bedau and Humphreys, 2008b; Rasmussen *et al.*, 2008].

The classic case of dynamic agency is the creation of a human. Figure 10.9 shows, in the sense of physiology, the hierarchy of a human from conception to birth to fully developed to death and beyond. The agency will vary as the layers vary in type and population. There is an up/down progression of hierarchical complexity from conception to birth and after death, when necrosis eliminates all cellular integrity and hence all agency at the cellular level (the hierarchy becomes a non-nested loose aggregation of atoms and molecules). As development proceeds, new hierarchy layers are added including tissue and complete organs and organ systems. At some point the number of organs stabilises although cell population sizes vary through the various developmental stages. Note that the prenatal environment $E(\cdot)$ is radically different to the postnatal environment $E(\cdot)$. Indeed the prenatal agency could technically be regarded as subsumed into the agency of the mother through being another organ of the mother. Remember the delineations we make are merely choices. Nevertheless the developmental process exhibits the characteristics of a change in a hierarchy and constitutes a form of dynamic agency.

Throughout the life cycle of the human over a range of timescales from milliseconds to years, the agency of the human is dynamic. Along with the hierarchy of the agency, the nervous system provides a human with another level of adaptation. This results in the social functioning of the human, who could be a politician or a rocket scientist. In any case the physical substrate upon which that adaptability is based is the (roughly) static[a] number of neurons and astrocytes of which the brain is constructed. Once the basic processes of development are completed, learning, of the kind captured in previous chapters, continually reorganises the cellular level of the brain and, to some extent, other organs and subsystems. In a comprehensive treatment of hierarchy, these

---

[a] In humans there is an early over-population of neurons that is then pared back during the development up to age 25-27. Thereafter the number of neurons is relatively static.

are the details necessary to capture the big picture of the human cognitive agency.

10.7 Emergence: Redux

Need I go any deeper into hierarchy than this? I can see an enormous potential for scientific exploration. However, for the basic purposes of this argument I am done. To introduce another descriptive aspect for science, all I needed to do was reveal the origins of causality and contrast it with what we scientists usually accept as a stand-in for it. I have put the observer inside the natural world and I have shown that the same causality gives rise to the observer. I have shown how the observer's embeddedness in the natural world creates a special relationship with both the observer's containing space and the ultimate collection of structural primitives of which everything is comprised. The same circumstances also reveal a connection (of as yet undefined nature/extent) between all entities in the universe. If I was a physicist, for example, right now I'd be thinking how obvious it is that entangled photons should remain so, and decohere together, even when hundreds of kilometres apart. The reason is that they are not actually hundreds of kilometres apart and are intimately connected through the underlying hierarchy at all times. Many things become obvious predictions when this kind of thinking is adopted.

But the most important thing about this is that in accessing actual causality, and describing the universe in the manner shown, I have not negated or refuted or in any way altered the validity of all existing 'laws of nature'. All I have provided is a way that the universe can be viewed, in which the *observer* is as explained as everything else, and indeed where we actually 'explain' in the technical sense or why/causality, for the first time.

What does the second-aspect tell us of the ghost in the machine? Well basically, it's found the ghost. Under the proposed second aspect, it is simply the real origins of causality. The second aspect also predicts that, to a single-aspect science of the old kind, all manner of incommensurable part/whole relationships are to be expected. They come about merely because the scientist intrinsically excises the part from the whole. In the dual aspect approach, the 'emergence' is literally the 'apparent' whole to an observer. The observer literally creates it. There never was any 'elan vital'. The elan vital is actually the fundamentals of causality itself, which is never actually contacted by single aspect science as practiced so far. As soon as you eliminate the observer, not by objectivity, but by *explaining the observer*, the big picture becomes clear. Therein we found our ghost.

> Examining the record of past research from the vantage point of contemporary historiography, the historian of science may be tempted to exclaim that when paradigms change the world itself changes with them. Led by a new paradigm, scientists adopt new instruments and look in new places. Even more important, during revolutions scientists see new and different things when looking with familiar instruments in places they have looked before. It is rather as if the professional community had been suddenly transported to another planet where familiar objects are seen in a different light and are joined by unfamiliar ones as well. Of course, nothing of quite that sort does occur: there is no geographical transplantation; outside the laboratory everyday affairs usually continue as before. Nevertheless, paradigm changes do cause scientists to see the world of their research-engagement differently. In so far as their only recourse to that world is through what they see and do, we may want to say that after a revolution scientists are responding to a different world.

[Kuhn and Hacking, 2012, Page 111]

In the dual aspect science chapter, the process of doing science in the second aspect will be made quite plain. It's in-principle straightforward. That being the case, what more has the history and discourse on hierarchy have to tell us? I have done a very large analysis of the history, and I find that it's a very interesting story, but that it won't tell us much more that has direct relevance to a scientific account of a scientific observer.

We can now leave this chapter having identified 'the underlying natural world' as the culprit behind causality and the reason for emergent properties. That there is an underlying natural world should be no surprise to anyone. Emanuel Kant (1724-1804), called it the 'noumenon' (in contrast to 'phenomenon'), and for various reasons lost in the vast philosophical burbling on the topic, it was never made a scientific target. The main reason for the lack of attention is the previously noted confusion between 'observation' (obviously via phenomena) and 'scientific evidence'. I am making it a scientific target for all the reasons in this book. I expect that huge numbers of scientists will fail to be able to see it. I also expect that those that can see it will form a very convincing lobby group. Especially when the dual-aspect science of the noumenon (the second-aspect) and phenomenon (the first-aspect) starts to bear fruit. We can now turn to that science.

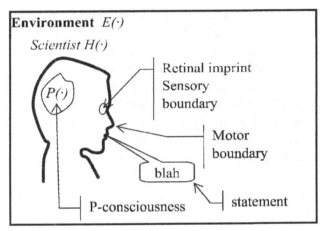

Figure 10.10   Using the *(·)* nomenclature in the 'what it actually is' depiction of the underlying structure of an embedded, embodied observer (scientist) inside an environment. The diagram is intended to convey the seamless unity of the underlying natural world. The nomenclature serves to delineate portions of the seamless unity.

10.8   Final preparation — establishing nomenclature

Consider Figure 10.10, which shows the diagrammatic form of the kind used in previous chapters. I have not properly explained the nomenclature yet. We now have enough background to do that. Figure 10.10 shows a human scientist denoted with a symbol, $H(·)$, embedded in an environment $E(·)$. The Figure 10.10 diagram looks straightforward, but I did not tell you the real intent of the *(·)* symbolism. I hinted at it in the explanation within Figure 4.2 when I said that the scientist is part of the environment (shown here as $E(·)$) in the way a human-shaped body of water is part of a lake of water. The reason for this is that what is depicted in the diagram is what the natural world is actually made of, not what it appears to be made of to us, within it. There is nothing in any scientific position ever put forward that has made the distinction. I am now making that distinction in Figure 10.10. What you are seeing is a single indivisible system of organised 'something'. I don't have to say what it is at this stage.

In this way I can contrast the 'what it is' and the existing 'what it looks like' ways of thinking about the natural world. This is necessary to be able to properly explain dual aspect science, the subject of the next chapter. The $(\cdot)$ nomenclature is from mathematics and depicts a mathematical function in a way that emphasises the sense of indivisibility of the depicted situation. Scientist $H(\cdot)$ is literally part of $E(\cdot)$ in the sense captured by the mathematical statement $E(\cdot) = G(H(\cdot))$, where $G(\cdot)$ captures the difference between $H(\cdot)$ and everything that is not $H(\cdot)$. This nomenclature needs no further development for our purposes here. It suffices to say that $G(\cdot)$ and $H(\cdot)$ are intimately related and become $E(\cdot)$ only when together, and that it is meaningless to speak about $H(\cdot)$ in any manner outside the context of the embeddedness of $H(\cdot)$ within $G(\cdot)$.

Within $H(\cdot)$ is a regional functional boundary that delineates the P-consciousness of $H(\cdot)$. It is denoted $P(\cdot)$. $P(\cdot)$ is part of $H(\cdot)$ in exactly the same way $H(\cdot)$ is part of $E(\cdot)$. These notional function boundaries are only that: a way of dividing up the natural world that does not recognise any separateness other than that associated with a functional or processing role. Such boundaries are important in $H(\cdot)$. There is the sensory boundary through which $P(\cdot)$ gets signals from the environment. In Figure 10.10 this is shown as the retina, but could actually be every surface that has sensory nervous apparatus. There is also the motor boundary. This is, in effect, what appears to us to be the external surface of the body that is actuated to cause behaviour. In Figure 10.10 this boundary is being depicted as causing $H(\cdot)$ to make the statement "*blah*". $P(\cdot)$ is shown the way it is because we know, from decades of neuroscience, that P-consciousness is a cranial central nervous system product. Now we know that with technical specificity. 100 years ago this was definitely not known. In the time of Kuhn's *Structure* it was also not solid science. It is relatively recent knowledge (as per Chapter 3).

This depiction of actual/underlying reality with the parentheses *(·)* should be thought of as unlike anything revealed by science so far. When we P-conscious scientists look at the natural world we do not see what is being captured with the bracketed terms. Instead, we see cells, atoms, trees, rocks and brains. But these are merely appearances that we cannot claim to be literal reality. To reiterate: the methods of empirical science enacted under the t_A/t_n framework cannot be claimed to access the underlying reality. Nor do they prove that the underlying reality is not scientifically accessible. Figure 10.10 depicts (1) the underlying reality of the natural world, and (2) the natural world as it appears, can be different things while all the scientific statements found to date remain perfectly valid. By choosing to treat the natural world this way, we get to examine the plausibility of a scientific world-view which recognises the difference between the world of appearances and the world of underlying structure.

Remember that the Figure 10.10 diagram depicts the actual underlying reality generating P-consciousness. What we 'see' (*with* our own P-consciousness) when we surgically open up a cranium is the appearance of that process: a brain in the act of delivering a first-person perspective (appearances) to the viewed brain. We do not see the P-consciousness delivered by the brain. By choosing to separate out appearances from underlying structure, the mysterious presentation of observational evidence of P-consciousness is now gone. By shifting our perspective to another kind of description – of an underlying reality that exists prior to all appearances - we are able to point to something that makes a first- person perspective happen while not invalidating the third-person (objective) perspective that is acquired and mediated by having that first-person perspective. It becomes possible to envisage a way of describing the origins of a first person perspective (P-consciousness) (a) without delivering it literally, (b) without invalidating anything objectivity says about how the reality looks to an observer within it and (c) without having to invoke bizarre principles like the mind-brain identity theorem to mask the obvious difference between appearance and structure. This can be done in a way that relieves objectivity-based behaviour of the responsibility for accounting for P-consciousness – which is also consistent with our measurement of (single-aspect) scientific behaviour – that told us to expect explaining P-consciousness to be impossible. This way of handling cognition and consciousness has separated and disentangled the traditionally paradoxical circumstances into three separate areas.

(1) An underling reality.
(2) A description of underlying reality (1) that accounts for how observation works within it.
(3) A description of how the reality appears when the (2) observation is used to describe (1).

This way of thinking also tells us several fundamentals:

- None of the (2) descriptions deliver 'what it is like'.
- None of the (3) descriptions deliver 'what it is like'.
- Only being a part of (1) actually delivers the experiences that constitute observation.
- Descriptions set (3) is not the reality (1), but *about* (1)'s appearances. This includes the $t_o/t_n/T/H$ framework that science uses.
- Descriptions set (2) is not the reality (1), but *about* (1)'s structure. We simply don't have a framework for the statements it might produce (yet).

10.9 Cognitive agency

Now we can elaborate a version of Figure 10.10 depicting a human as a form of cognitive agency in a more technically robust way. This is shown in Figure 10.11. I have recognised the environment as being *situated* within the universe $U(\cdot)$ within which is *situated* the aforementioned environment $E(\cdot)$ in which cognitive agency $H(\cdot)$ is *situated*. Situated agency is a technical term with a large literature, and it is meant in exactly the technical sense of that literature. Note also that Figure 10.11 is consistent with the cultural learning structure depicted in Figure 7.2.

Within $H(\cdot)$ is *Brain($\cdot$)* that we know generates phenomenal consciousness $P(\cdot)$. $P(\cdot)$ is now depicted as interacting with some other portion of *Brain($\cdot$)* and as a result of their mutual interaction, a belief b_n is formed. Note that the separation of $P(\cdot)$ and *Belief($\cdot$)* does not imply their physical separateness. The two systems can be physically collocated. The symbol b_n does not have the $(\cdot)$ form and does not signify any particular portion of the natural world. It merely serves to label a particular brain configuration (or better, a trajectory) that expresses b_n within *Belief($\cdot$)* that contributes to the operation of what is labelled the boundary of $H(\cdot)$. The belief b_n might contribute to an ability to play tennis, in which case it is a form of 'tacit knowledge'. Another form of tacit knowledge is an ability to speak, in which case, b_n might be a scientific statement that impacts the boundary of $H(\cdot)$ in such a way that it is conveyed to another scientist.

The manner in which $H(\cdot)$ operates is via the boundary through the subsets of $H(\cdot)$ labelled *Sensory($\cdot$)* and *Motor($\cdot$)*, whose function is obvious. Beyond the boundary of $H(\cdot)$ is shown environmental impact systems $S(\cdot)$. These are not part of $H(\cdot)$, but are part of $E(\cdot)$. They may be

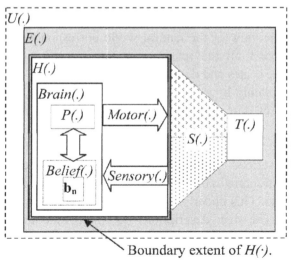

[P($\cdot$)] denotes 'what it is like', the actual experience of being the subset $P(\cdot)$.

Boundary extent of $H(\cdot)$.

Figure 10.11. Situated cognition and the underlying reality fully expressed in the $(\cdot)$ nomenclature including the $[\cdot]$ 'first-person' operator.

the air transmitting sound waves or perhaps sunshine on the surface of $H(\cdot)$, some of which reaches the retina and other parts of which feel warm because $P(\cdot)$ delivers that experience through the action of *Sensory(·)*. Operating in the reverse direction is the impact that $H(\cdot)$ has on $E(\cdot)$ through *Motor(·)*. Collectively these environmental influences actually surround and encapsulate $H(\cdot)$, but are shown as $S(\cdot)$. Note that $S(\cdot)$ actually selects a subset of all environmental influences that relate to an object, also within $E(\cdot)$ with which $H(\cdot)$ is interacting, labelled $T(\cdot)$. The final nomenclature is $[X(\cdot)]$, which is used to denote the 1PP of the subset $X(\cdot)$. This is as developed as the nomenclature needs to be for the purposes of establishing dual-aspect science practicalities.

10.10 Summary

This chapter used the obvious lack of causality (in present scientific statements) as a way to extend science into a form in which handling causality is a natural expectation. The method examined the difference between the world as described by current science (under the $t_A/t_n/T/H$ framework), and an actual (underlying) natural world. To clearly reveal the difference between these two things, an analysis of hierarchy was used. This second scientific view of the natural world corresponds to a trans-disciplinary view across all the sciences, tracing structure from macroscopic to microscopic scales and deeper.

The analysis revealed plainly how the world of the existing $t_A/t_n/T/H$ framework is a collection of descriptions, acting along an 'aggregation axis', by a presupposed observer. The analysis then revealed the possibility of a separate description of the natural world, of an underlying hierarchy of yet-to-be-described structural primitives that provide a separate account along a 'structure axis' that traverses orthogonally, from microscopic to macroscopic. This underlying natural world was revealed as an indivisible whole. The new descriptions do not invalidate the existing ones. They are a new kind of scientific statement about the same natural world, describing its intrinsic unfolding as it actually is, prior to any observer. Indeed it is the job of this underlying unity of structural primitives to 'unfold' the scientific observer that sees the world, scientifically, like humans do. This 'second-aspect' science therefore cannot be a product of the human observer (mind).

What we are missing is a way of exploring this underlying natural indivisible whole.

Chapter 11

Dual Aspect Science

History of science indicates that, particularly in the early stages of development of a new paradigm, it is not even very difficult to invent such alternates. But that invention of alternates is just what scientists seldom undertake except during the pre-paradigm stage of the science's development and at very special occasions during its subsequent evolution. So long as the tools a paradigm supplies continue to prove capable of solving the problems it defines, science moves fastest and penetrates most deeply through confident employment of those tools. The reason is clear. As in manufacture so in science – retooling is an extravagance to be reserved for the occasion that demands it. The significance of crises is the indication they provide that an occasion for retooling has arrived.

[Kuhn and Hacking, 2012, Page 76]

The previous chapters were spent revealing one of Kuhn's 'very special occasions'. Time to retool. The concepts and some portions of the following text were first published in the Journal of Consciousness Studies, vol. 16 (2009), pp. 30-73.

11.1 Motivating a new science framework

In the absence of formal self-governance (in contrast to self-regulation - see Chapter 1), it seems reasonable to consider that each of us, as individual scientists, has the right and duty to attend to our own self-governance, including questioning beliefs inherited from scientific forebears. The beliefs in question are those implicit in the behaviour of mentors, which we now know confines us, through the imitative learning process, to creating and managing a population of scientific statements, each of kind t_n, defined in equation (8.2) as

$$t_n = \quad \text{The natural world in} < \textit{insert context} > \qquad (11.1)$$

behaves as follows: < *insert behaviour* >

These statements are aggregated into a set T of accepted statements, and a set H of rejected/hypothesised statements as follows:

$$T = \quad \{t_A, t_1, t_2, \ldots, t_n, \ldots t_{N-1}, t_N\} \tag{11.2}$$
Accepted

$$H = \quad \{h_1, h_2, \ldots, h_m, \ldots h_{M-1}, h_M\} \tag{11.3}$$
Rejected, Hypothesised

The statement t_A is special statement (Chapter 8 equation (8.4)) that describes the natural world (humans) when it is creating scientific statements, and it too is of the form t_n:

$t_A =$ The natural world in < *the context of a human*   (11.4)
being scientific about the natural world > behaves
as follows: < *to create and manage the members*
of a set of statements of type t_n, each of which is a
statement predictive of a natural regularity in a
specific context in the natural world external to
and independent of the scientist arrived at
through the process of critical argument and that
in principle can be refuted through the process of
experiencing evidence *of the regularity*>

This behaviour does not explicitly refer to sets T or H. All equation (11.4) does is say that set members will be created, modified or removed from a set. A future version of equation (11.4) might explicitly mention sets T and H. The above equations are only a first-pass suited to my purposes here.

We have already seen how a process of scientific behaviour defined by equations (11.1)...(11.4) places us implicitly within an environment of predictable permanent failure to account for the scientific observer/P-consciousness. This position has real ramifications. This is not some esoteric argument about abstract theoretical positions. This is about behavioural options available at the real scientific coal-face. If you are in neuroscience, what is at stake is literally the failure to fully understand how anesthesia works or exactly what chronic pain is. Ask the sufferers of chronic pain if we scientists should refrain from certain behavioural options even a millisecond longer than we need, given that limitation is actually causing a failure to fully understand their affliction. This is a rather obvious emotive argument, but it highlights our responsibility as individual scientists to make sure that if we restrict our options merely through implicit unchallenged assumptions delivered into our behaviour by imitation, then we deserve to be held accountable and to have them flushed out and made explicitly obvious as such. It is what we expect everywhere else in science. What form of special exemption can we claim unique to ourselves? This is enough to motivate me. Perhaps it will speak to you as well.

Previous chapters showed us confined to an assumed scientific behaviour, t_A, that accounts for everything except the scientific observer, which is another way of saying that a scientific account of P-consciousness is outside the scope of our behaviour. Our lack of attention to explicit management of t_A is an implicit assumption that scientific behaviour is complete. Our behaviour, t_A, is all there currently is to scientific behaviour. Logically this entails that the scientific behaviour that will succeed is something other than t_A. This makes our assumption obvious: *we are assuming that scientific behaviour itself is developmentally completed.* The solution is thus very simple: *we must complete the development of scientific behaviour* by adding a new scientific behaviour into set T. We already have behaviour t_A, so let's call the new behaviour t_S. It creates a new kind of scientific statement. Let us be consistent and assume t_S can be added to set T by using t_A in the prescribed way. This means that set T, if the upgraded science framework becomes accepted, will have the following form:

$$T = \{t_A, t_S, t_1, t_2, \ldots, t_n, \ldots t_{N-1}, t_N\} \qquad (11.5)$$

Let us also recognise that the behaviour t_S, whatever it is, does not involve presupposing the observer (a human). It creates statements that are very different to set T, and that do not depict the results of observation ('organise appearances'), yet are at least as revealing of the natural world as set T. We need somewhere to keep these new statements, so by analogy with what we already have, let's define new sets T' and H' as follows:

$$T' = \{ t'_1, t'_2, \dots, t'_j, \dots t_{J-1}, t_J \} \qquad (11.6)$$
Accepted

$$H' = \{ h'_1, h'_2, \dots, h'_k, \dots h'_{K-1}, h'_K \} \qquad (11.7)$$
Rejected, Hypothesised

For those readers with a more mathematical bent, the existence of this new set T' implies that the entire set of human scientific statements would become a set union:

$$\{T_{TOT}\} = \{T\} \cup \{T'\} \qquad (11.8)$$

From equations (11.5) and (11.6) we can immediately expect some kind of operational equivalence between generic statements t_n and t'_j. Generic statement t_n is expressive enough to allow its adaptation to set T'. No new definition is needed. According to equation (11.5), a human can now behave scientifically in two distinct ways, both of which are in set T as self-consistent, self referential scientific statements describing the natural world when it is behaving scientifically. The only remaining unknown is t_S in set T, which tells us what the new behaviour actually is.

I must stress that the contents of sets T and T' are identically treated as statements. As two different representations of the same natural world, the statements may have a different form. Yet they are accepted merely as human utterances in the sense of the previous discussions. Just like set T, there is no claim that set T' contains 'facts' or 'theories' or 'laws' or 'truths'. Accepting a statement into a set imputes no status on the statement in any of these terms. Regardless of their form and content, they are to be treated as tokens exported by one brain to be used by another with the intent of eliciting consistent behaviour in each. One such behaviour is that of scientific prediction. If you must have a classification, then think of sets T and T' as *'usefully predictive'* and sets H and H' as *'not yet usefully predictive'*.

11.2 What is the fundamental character of the contents of set T'?

Previous chapters tell us that set T' does not describe the natural world like set T does. Set T is to description (what) as set T' is to explanation (why). We have seen already that set T is devoid of causality – that which necessitates that the world unfolds the way it does. Therefore, set T' shall somehow deliver a description of the causal necessity (mechanism) that makes the natural world unfold as it does, *including the unfolding of the scientific observer that sees the world as per set T*. If that is to be the case, then set T' must also somehow directly address whatever it is that is doing the unfolding. Whatever that is, it is not the subject of any set T statement. Set T' has no presupposition of the observer, by definition. We are creating set T' to account for the observer. In accounting for the observer set T' also intrinsically connects to its scientific evidence, not by any particular observation, but through the existence of observation at all.

One or more of the standard statements in T' must form an account of the mechanism that gives rise to an ability to observe at all. T' contains a very different kind of statement: the causality of the scientific observer is within it, and is delivered as a description of the unfolding of as-yet-unspecified building blocks. T' does not describe the appearance of a natural thing. It delivers a description of the actual underlying structure of the thing. Let us call the statements in set T 'appearance-aspect' and those in set T' 'structure-aspect'. Together they give this chapter its name: dual-aspect science (DAS). This categorisation has been pre-empted in the statements t_A and t_S, as statements that depict the human behaviour that results in appearance-aspect and structure-aspect statements, respectively.

11.3 What is the behaviour t_S that populates set T'?

The structure-aspect can use the appearance-aspect generic statements. For the sake of showing how t'_j, may differ from t_n, consider

$t'_j =$ The natural world in the context of < *a particular*   (11. 9)
configuration   of   explored   structure-aspect
descriptions > behaves as follows: < *insert the*
results of the structure-aspect behaviour here>

This might be thought viable in some later formalisation of dual-aspect science by others. However, there is no obvious necessity for a fundamentally new form for t'_j. Based on the essential properties we know it must have, the simplest generic form of the behaviour, t_S, that populates set T', is as follows:

$t_S =$ The natural world in < *in the context of a human*   (11. 10)
operating in the structure-aspect description of
the natural world > behaves as follows: < *to*
construct statements of type t'_j, *of the same basic*
form as t_n, *that populate a set of such statements*
T' *and that define the initial conditions and*
interrelationships of instances of an hypothesised
structural primitive which shall be entirely
consistent   with   the   'appearance-aspect'
statements contained in set T by provision of the
actual mechanism for creation of an entity
capable of populating set T >

We have now completed the entire DAS framework in the most generic way possible, without defining any practicalities, specific examples or its detailed empirical justification. The above proposition is entirely empirical, although it may not look it just now. At this stage consider t_S as a hypothesis in set H ready for testing to see if it makes it into set T.

11.4 Empirical implementation

Without specifying the details of a particular structure-aspect proposal, we can assert from the start that the dual-aspect framework does not merely get sanctioned by agreement. It has to empirically earn its position as a valid framework. The practical process can be established as an empirical test on another set H hypothesis; a test to see if a new scientific behaviour, t_S, can be added to the existing behaviour t_A. Formally we may consider the establishment of DAS through proving the following statement is predictive of the natural world of human science:

$h_{DAS} =$ The natural world in < *the context of a*   (11.11)
human being scientific about the natural world > behaves as follows: < *to construct two completely separate but intimately related sets of scientific statements, the first of which populate the existing set T, which becomes the appearance-aspect, and the second the structure-aspect (as outlined in document X), which shall populate a set denoted T′ and in which both sets T and T′ are empirically validated through the explanation and subsequent use of the human faculty of P-consciousness to enact scientific behaviour.*>

This is an interim hypothesis designed to allow the exploration and potential implementation of dual-aspect science starting from a position of single-aspect. In its practical execution the statement h_{DAS} demands that structure-aspect scientists (physicists) make appearance-aspect predictions of brain material (neuroscience) appearances consistent with an explicit mechanism for delivery of P-consciousness by the chosen structural primitive.

Once that is done, then the statement t_S in set T becomes accepted as normal scientific behaviour, and set $T′$ joins set T as a repository of the (as yet unspecified) new kind of statements. The DAS framework thereby becomes the norm for scientific behaviour. This is an unusual circumstance and it delivers another reason why the process has been hidden for a long time. Empirically testing h_{DAS} demands an unprecedented linkage of two historically very separate disciplines: physics and neuroscience. Not only that, the kind of physics that is being done is structure-aspect and is not something that the great bulk of physicists will recognise (even though some of them do it already – to be discussed later). This unusual connection is shown in Figure 11.1 (h_{DAS}), where T' joins T in dual-aspect science. For every appearance-aspect account in T there is a corresponding structural aspect account in T'. As a result of the cross-linking, all scientific statements are put on a more rigorous footing.

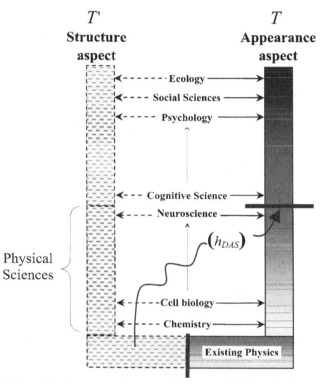

Figure 11.1   Dual aspect science and its empirical validation via the physics-neuroscience collaboration in explanation of the scientific observer.

The structure-aspect literally computes a representation of the underlying universe. It computes the structural hierarchy of the natural world as described in Chapter 10. The computation itself is literally the structure-aspect scientific 'statement'. It could never have been done in the past, but now we have computing of the kind that can begin to tackle it. In addition we can look at what general principles might apply to all structure-aspect activity including any general principles behind an observation mechanism.

Appearance-aspect science accumulates a hierarchical form of knowledge in set *T*. In Figure 11.1 set *T* statements are associated with all the obvious horizontally listed science subdisciplines. A test for h_{DAS} uses a structure-aspect physics result to make an appearance-aspect prediction in neuroscience, where it acts to explain the neuroscience of the scientific observer via the Figure 11.1 (h_{DAS}) route. This is no simple feat. However it has logical teeth. A structure-aspect scientist with a novel set *T'* membership that corresponds to brain material set *T* appearances expressed in detailed molecular and cellular biology terms is in a very commanding logical position in support of h_{DAS}. Until we actually start thinking along these lines, and properly analyse the complexity levels and look for shortcuts, nobody is in a position to say how viable it is. It certainly does not seem as intractable as, say, the human genome project once did.

At the same time it is ironic to see that the explicit empirical acceptance of a similar statement h_{SAS} in support of confinement to single-aspect science has never been carried out. We are simply acting as if it is true, living with the consequences of its implicit presence in our scientific lives, and passing on the behaviour t_A by imitation. No empirical or theoretical analysis ever carried out by science has proved that statement h_{SAS} captures the only possible scientific behaviour. The reason it is currently the only scientific behaviour is because it is the only one we actually do.

11.5 The DAS framework: Overview

As advised in the previous chapter, when you take an overview of the DAS situation, you find yourself involved with three fundamental things:

(a) An actual underlying universe, $U(\cdot)$, made of some kind of structural primitive(s) relating to each other in specific, regular ways.

(b) A scientist within $U(\cdot)$, made of the same structural primitives $U(\cdot)$ is made of, populating a set *T'* with abstractions of the structural primitives and their rules of interaction in such a way as to produce a representation of the persistent structure we call a scientist, complete with an observational faculty called P-consciousness. This is the *structure-aspect*.

(c) The same (b) scientist also populates a set T with abstractions predictive of the way $U(\cdot)$ appears to the scientist when $U(\cdot)$ is observed using the P-consciousness supplied by processes revealed in (b). This is the *appearance-aspect*.

Appearance-aspect science (c) is not literally the natural world, but merely about it. Structure-aspect science (b) is likewise not the natural world, but merely about it. Currently we do (c) alone. The revised framework is posited to be 'as good as it gets' (technically maximal information content) when it comes to understanding $U(\cdot)$ from within. The actual structural primitives comprising $U(\cdot)$ are otherwise admitted as intrinsically unavailable to us in exactly the same way as a human-shaped collection of water in an ocean cannot possibly know water because collections of behaving water are used to observe the ocean, and in the creation of the representation, the original water is gone. The painting cannot view the painting. You can't telephone someone a telephone. And so forth. Metaphors and analogies abound.

The simplest kind of $U(\cdot)$ is one made of a large collection of identical structural primitives (as examined in the chapter on hierarchy). This is the $U(\cdot)$ that is recommended as a starting place in T' science and the one that I have actually been exploring for some time. No matter how successful or otherwise my results may be, they are invalid unless a framework exists in which they are meaningfully expressed and discussed. That framework is the dual-aspect science framework outlined above. One crucial and striking feature of the framework, that must be remembered above all else, is that it is $U(\cdot)$ (item (a) above) that delivers P-consciousness. As such an explanation of P-consciousness (indeed explanation generally) is actually contained in set T', not set T. The dual-aspect science framework thereby makes it clear that the appearance-aspect (c) alone was never ever able to predict or explain P-consciousness, and expectations that it could are mistaken. But I am ahead of myself. We have not yet detailed how to use the new framework. So it's time to do that.

11.6 Practical exploration of set T': The natural cellular automaton

I propose one way of exploring the structure-aspect as follows: We know that the generic member of set T', t'_j, is a depiction of a collection of a known number, say N'_j, of a chosen (identical, as suggested) structural primitive, say X'_j. We can consider that each can be in a number of states that define the basic primitive. Each structural primitive is established in an initial state, say I'_j. Each structural primitive interacts in some way with the other members of the collection. Usually this means that some subset of the collection becomes a local neighbourhood for each

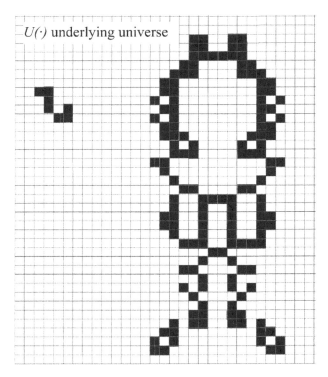

Figure 11.2. A basic 2D cellular automaton. Structural primitives are cells that can be black or white. The rules of interaction between structural primitives is $R'_j = $ *If a structural primitive is in state 'black', its next state will be 'white' if the number of 'white' flanking states is less than 2. Otherwise its next state will be 'black'.* 'Flanking' defines a neighbourhood of up to 8 cells for each cell.

structural primitive. We can define these neighbour interactions according to a set of rules, say R'_j. The exploration is done computationally. Each new set T' member is a variant of N'_j, X'_j, I'_j and R'_j, sampled and frozen at some revealing moment in its computation. It may be explored for multiple initial conditions I'_j. It's that simple.

Notice the sudden change in terminology. I am not talking about a result in terms of it being an 'observation' of, say, Thing-X. The computation *is* literally a description of the underlying world expressing Thing-X. It is not something that came from a human mind. The human mind initially establishes N'_j, X'_j, I'_j and R'_j. Computers then deliver the actual statement.

Thing-X, inextricably and seamlessly enmeshed with its computed environment, is evident in emergent behaviours inside the computation itself. What the computation expresses can be sliced and diced with computational probes to detect properties that might be space, atoms, fields, Thing-X and so forth. As human observers, these are the objects we expect to find inside such a computation. But they are not observed, they are computationally delineated. We then observe the delineation. This sounds like an onerous computational task, but until we actually start we can't possibly know if there are ways to streamline the process.

Once we identify such things as atoms and electromagnetic fields and mass, we can start to explore the relationship between any of these things and the computed surrounding environment. It is this property that our brains exploit. Somehow, in the mass of numbers, there is a way in which an observer (X) can let their own state be altered by points distant (NOT-X). This is the real mystery that appearance-aspect science cannot touch. Once we can see that potential relationship within the computation, then we can start to see how a brain might be taking advantage of it in the real thing. When we find that we have done the job.

N'_j, X'_j, I'_j and R'_j are precisely the general form of the 'cellular automaton' (CA). The CA is a well travelled mathematical formalism and the perfect vehicle for an exploration of the structure-aspect. There are many online CA videos complete with explanations. Search for 'Cellular automata game of life'. Figure 11.2 shows a primitive CA. It is a 2-dimensional CA, and because it is 2-dimensional it is possible to show it entirely on a single flat page. The real thing will be more sophisticated, but the structure-aspect CA is conceptually very similar. It has all the basic properties of N'_j, X'_j, R'_j and initial conditions I'_j.

In the Figure 11.2 case the universe $U(\cdot)$ is a presupposed 31 x 39 grid of 'structural primitives' ($N'_j = 1209$) represented by a 'cell' that can have either a black (B) or white (W) state $X'_j = \{B,W\}$. The initial state of all the structural primitives is established as I'_j, a list 1209 cell states, one for each cell such as $I'_j = \{B,B,B,W,B,W,W,W,\ldots.etc\}$. The rules of interaction for structural primitives, R'_j, involve a simple counting process detailed in the Figure 11.2 caption. Note the abstract appearance of the CA. This is to be expected in the structure-aspect. There are no appearances. You are inspecting a computational space, not a real space. The computational substrate that implements the structure is implicit. Each cell is inspected, and its next state is computed based on the current state and the state of relevant neighbours. Then the entire grid is updated to the computed next states for each cell. This process happens repeatedly for as long as necessary.

It is well known that simple computational rules like this can lead to intrinsically unpredictable states and even total randomness, in spite of the entirely fixed (deterministic) rules. The cellular automaton is not describing how $U(\cdot)$ appears, it is actually computing a representation of the underlying universe itself. Figure 11.2 is a little deceptive in that the drawing feeds our innate tendency to consider it as having spatial extent. In more realistic CA space is also an emergent property (as per Chapter 10) and the relationship of the structural primitives and the recognisable things of common experience (the appearance-aspect) are all emergent properties of subsets of primitives made visible by the CA itself.

Note that the succession of states also has the apparent status of what we would call time in the appearance-aspect, thereby providing the basis for how it all appears to us (as shown in the chapter on dynamical systems). A formal representation of the state of particular structure-aspect science would be to install a new t'_j in set T' that references all the specifics of N'_j, X'_j, R'_j, initial conditions I'_j, along with any other technical specifications for the CA. Then, the outcome's particular computational instances are reported, along with their interpretation. That is structure-aspect science in practice.

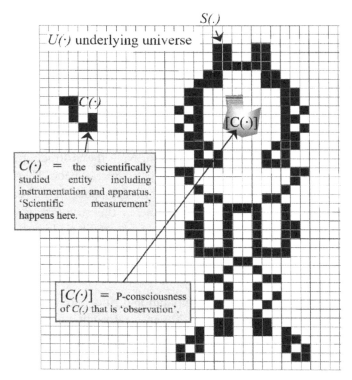

Figure 11.3.　P-consciousness, appearance and the structure-aspect as revealed in a CA. Appearance-aspect Scientist $S(\cdot)$ studies $C(\cdot)$ using its appearance $[C(\cdot)]$. The structure-aspect science of that situation is literally the CA. What has traditionally been called scientific observation is literally $[C(\cdot)]$. Now we have another aspect, scientific observation can also be classed as the act of inspecting the CA.

11.7　P-consciousness and the CA

We now take a look at both aspects and the proving of h_{DAS} from the perspective of a CA. Figure 11.2 only reveals numbers resulting from a computation (a particular structure-aspect t'_j) that depicts a natural world made of a certain structural primitive. We know that for a chosen structural primitive to remain in set T' the t'_j must be capable of expressing sufficient complexity to create an observer/environment as per the h_{DAS} requirement. By taking the computational CA result and then by slicing, averaging and projection into our world of appearances, it must make predictions like atoms, space and so forth and, ultimately, observable neuroscience predictions that we human observers can find, that are recognised as part of the process inside t'_j that corresponds to the observer in the act of observing something in the computed observer's environment (also within t'_j).

Figure 11.3 shows Figure 11.2 interpreted to reveal how P-consciousness fits into the scheme. Figure 11.3 right hand side has been interpreted as a CA computed human $S(\cdot)$ observing a CA computed object $C(\cdot)$. The underlying grid reveals the computational signs of P-conscious experience $[C(\cdot)]$ of distal object $C(\cdot)$. This is revealed as the computed connection between $C(\cdot)$ and $S(\cdot)$ indicated by the dashed lines. Treated as a representation of an act of observation by $C(\cdot)$, and an understanding of P-consciousness that a much more sophisticated CA may give us, we know that to $S(\cdot)$, the first-person percept $[C(\cdot)]$ actually appears projected onto the real external object $C(\cdot)$. In this way the underlying real $C(\cdot)$ grid features are masked. $C(\cdot)$ is thereby immersed in a world of experiences (P-consciousness) and able to do appearance-aspect science.

We can now see the importance of the nomenclature system introduced in the previous chapters. The bracketing system $(\cdot)$ actually refers to the underlying structure-aspect computation. Note that if Figure 11.3 $S(\cdot)$ was a scientist studying $C(\cdot)$, then the first-person perception of the object, $[C(\cdot)]$, is the act of scientific observation by the scientist $S(\cdot)$ in appearance-aspect science enacted through behaviour t_A. As a result of the behaviour of $[C(\cdot)]$, $S(\cdot)$ populates a set T with an appearance-aspect statement t_n. The structure-aspect equivalent to the appearance-aspect science is literally the CA itself. To further reinforce the disparity between the CA depiction of observation by $S(\cdot)$ and the actual experience of $S(\cdot)$, from the first-person perspective, Figure 11.4 depicts how $S(\cdot)$ and $C(\cdot)$ appear to the scientist outside Figure 11.3 that generated Figure 11.3. The differences between the underlying structure and the appearance of it have been exaggerated.

I hope you now have a better appreciation of how the DAS proposition becomes qualified through the h_{DAS} verification process. You should now be able to understand how the perceptual (P-consciousness) world of $S(\cdot)$ masks the reality of the underlying structure and the rules that drive it from one state to the next. This masking process results in appearance-aspect science. In adopting a dual-aspect science approach, however, scientist $S(\cdot)$ also gets to make CA propositions like Figure 11.3. This explicit understanding of our situatedness shows us the underlying principle of dual-aspect science: *we are inside the system we seek to explain, made of it and operating as scientists because of the facilities provided by that circumstance.* Because we are inside it, made of it, we get two ways to depict it to ourselves. I could speculate about the concept of being entirely outside $U(\cdot)$, but this is probably a meaningless concept, and from that perspective, the view of $U(\cdot)$ would seem, to me, to be rather impoverished. Only we inside get 'what it is like'.

In the real natural $U(\cdot)$ that is our universe, the 'grid' or 'cells' do not exist separately. Nor is it 'computed' on anything. To effect this reality in structure-aspect science, the computed cellular automaton must be chosen appropriately. Each structure-aspect CA 'cell' is a computed dynamic structural primitive interacting with neighbours using rules of affinity for neighbours defined by us, commencing at some kind of initial

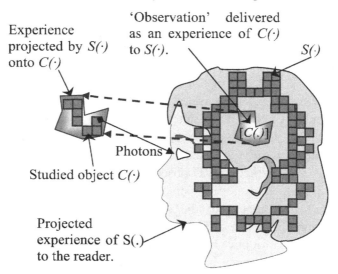

Figure 11.4.   This is how Figure 11.3 appears to the scientist that generated it. The reader's subjective view of $S(\cdot)$ is superposed.

conditions also defined by us. We must implement this computation in a way that reflects the computer-less reality, $U(\cdot)$, that it is depicting, so that localised persistent structure emerges naturally and does not include artefacts related to our use of computers. Each computed structural primitive must change state when local conditions (neighbours) permit. The results of the computation will be confusing to look at. What we call space is just as 'computed' as everything else (e.g. what we call matter), and all are comprised of the same set of primitive structural element(s) hierarchically organised as previously discussed. There will probably be fundamental properties of systems like this which we will eventually come to understand give rise to the possibility of P-consciousness. Exactly what these fundamental properties are will not be accessible until structure-aspect science commences.

11.8   More on the structure/appearance divide

Here we consolidate our idea of scientific behaviour as the behaviour of a natural hierarchy (the scientist) that reveals the natural hierarchy in the form of scientific statements about the various Figure 11.1 layers. Consider Figure 11.5, which depicts a scientist $S(\cdot)$ facing an abstracted hierarchy, of which $S(\cdot)$ is an instance. Deep down in the hierarchy we get the scientific statements of physics. High up the hierarchy we get the scientific statements of social sciences, ecology and so forth. Consider that $S(\cdot)$ meets the hierarchy within a shared environment, $E(\cdot)$, also a natural hierarchy, say, Earth, which is taken to be at organisational level (N+i). Meanwhile, scientist $S(\cdot)$ studies the hierarchy at and around layer N, say the cellular level within brain tissue, at which it is believed P-consciousness arises and thereby enables $S(\cdot)$ to do science as described here.

We have already encountered the reality that, no matter how many sub-sub-sub-layers there are, both $S(\cdot)$ and the studied hierarchy are *all of it* (down to Figure 10.2 layer Ω). In Figure 11.5 the bottom of the hierarchy, the starting layer that the most ideal form of structure-aspect science would address, is at layer (N-j = Ω). We have also encountered the reality that the hierarchy is a single containment hierarchy that contains itself. We know that the highest level of containment (N+k), and the set of all of the tiniest sub-division of $U(\cdot)$, layer (N-j = Ω), are identities. In terms of Figure 11.5, $U(\cdot)$ can be divided and divided until it reaches a structural primitive. That is layer (N-j = Ω). The collection of all such primitives is layer (N+k), also signified as $U(\cdot)$.

What is actually depicted in Figure 11.5 is an act of appearance-aspect science that involves the meeting of two hierarchy layers through the shared subset $E_N(\cdot)$ of environment $E(\cdot)$. Because of the behaviour of $S(\cdot)$, layer N is currently under scrutiny. $S(\cdot)$ is doing cell biology. This is the way the appearance-aspect would depict the connection of two layers in the hierarchy. Humans cannot see cells without technology. Therefore $E_N(\cdot)$ includes such things as a microscope. The net effect of $S(\cdot)$ studying layer N is to establish a layer of causal relations between layer N and $S(\cdot)$, resulting in scientific observation, which is signified, as is usual here, as happening within $P(\cdot)$, a subset of brain $B(\cdot)$. Using the

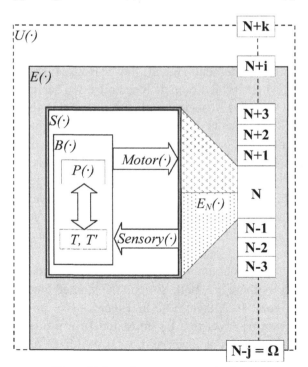

Figure 11.5.   Appearance aspect science.

nomenclature established previously, we would say that the subjective experience of layer N is [$P(\cdot)$], where the entire chain of causality from layer N through $E_N(\cdot)$, *Sensory*($\cdot$), $B(\cdot)$, *Motor*($\cdot$), $E_N(\cdot)$ and back to layer N has participated in the production of a first-person perspective (1PP).

Where the appearance-aspect reveals to us a view *across* the hierarchy at a given layer level, the structure-aspect depicts a route *down* the hierarchy. The structure-aspect depicts the vertical accretion of connected structural primitives that expresses structure from the very small spatiotemporal domain/scales to ever larger spatiotemporal domain/scales. When observed this structure is made visible, to an entity made of the same hierarchy, by contact across the hierarchy, at each layer along the lineage (N-j = Ω) ... (N-3) ... N ... (N+3) ... (N+i) ... (N+k). In this way the structure-aspect is, in a sense, *orthogonal* (at right-angles to) to the appearance-aspect. Causality is actually enacted from the lowest layer to the highest, and some causality can then act across between different layers of organisation. This is the looping nature of causality revealed in the chapter on hierarchy.

To see this orthogonality more directly consider Figure 11.6 where the two axes implicit in Figure 11.5 are made more explicit. Consider the organised gray array underneath to be a computed structure-aspect result. Let it be organised, say, in terms of the computer memory storing the result. By inspecting the memory and the relations between bits of it in terms of its depiction of structural primitives, we get to explore the structure-aspect in which we see an organisational hierarchy of interacting structural primitives. These organisations we analyse and find that parts of it correspond to space and other parts to atoms, molecules and so forth. Some of the structural hierarchy is not obviously categorisable, and might fit as part of the recognisable layer above, the more recognisable layer below, or neither layer. The fact that appearance-aspect science renders such intermediate structure as invisible is simply the way it is. In its mature state, dual-aspect science will explicitly account for why that is the case.

The disparity between the *T*-aspect and the *T'*-aspect is made more evident in Figure 11.6 in that when the structure-aspect is carved into neat accretion layers, it collects groups of clearly delineated structural primitives (square boxes). On the other hand, those collections of structural primitives revealed by the T'-aspect do not necessarily coincide with the T-aspect. These aggregations are shown by the superimposed irregular 'rectangles'. Both aspects account for the entirety of the hierarchy, but in principle, T can only partially reveal T'.

The depiction of the gray substrate as a computer output renders both T and T' visible. In that depiction one can see the structure axis (down the Figure 11.5 hierarchy) as being at ever-finer granularity with smaller and smaller boxes. What is not shown is how two entities interact (such as $S(\cdot)$ and layer N in Figure 11.5).

At this point it might be instructive to go back and review the looping/circular causality as it is depicted in Figure 10.6. With a little mental rearrangement, the diagram depicts the causality of the scientist⇔studied object relationship. Look at the entire system of causality. First there is the 'aggregation' or 'appearance-aspect'

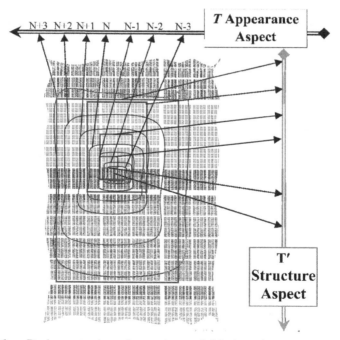

Figure 11.6.   Dual-aspect science in practice as visible from the neutral computational substrate (rectangular grid corresponding to structural primitives). Fixed nested subsets of the underlying grid form the structure-aspect descriptions. Horizontal axis appearance-aspect layer descriptions need not correspond exactly to the structure-aspect.

horizontal causation between 'the scientist' and 'the studied'. Now, in addition, we have a huge vertical 'structure axis'/structure-aspect causal lineage all the way down to lower layers, across and back up again. The scientist and the studied are obviously connected in unexplored ways. That connection, it is posited here, is a crucial component of P-consciousness. It is only visible through a structure-axis approach under the dual-aspect science framework.

11.9   The natural CA

Finally, consider the hand-constructed natural CA shown in Figure 11.7. It is of the kind 'network-CA' [Wolfram, 2002]. It is intended that the CA is literally depicting the notional circumstances of Figure 11.2 to Figure 11.5. In it we see scientist $S(\cdot)$ studying natural world $C(\cdot)$ inside universe $U(\cdot)$. Figure 11.7 CA 'nodes' are structural primitives that have an affinity for three and only three neighbours. There are 5368 structural primitives arranged in a grid that is actually a toroid. The top row wraps around to the bottom row. The left column wraps around to the right column. Each structural primitive has exactly three other cells to which it is connected. The connection to, say Figure 11.3, is notional and intended to show the basic principle. In Figure 11.7 the most important features are the fact that the entire thing is a single, seamless entity with structures within it, and that it represents the universe in a dimensionality higher than three, and in which space itself is a computed CA behaviour.

As observers $S(\cdot)$ inside Figure 11.3 we can look at $C(\cdot)$ and walk around it. In reality $C(\cdot)$ is internally connected to what we call space, and the space is hyper-connected in ways yet to be explored. As a result, the innards of $S(\cdot)$ and the innards of $C(\cdot)$ are actually connected and/or relate to each other in a computationally understandable, rigorous, persistent way. This conceptual adjustment underlies the approach to understanding how P-consciousness works in $S(\cdot)$.

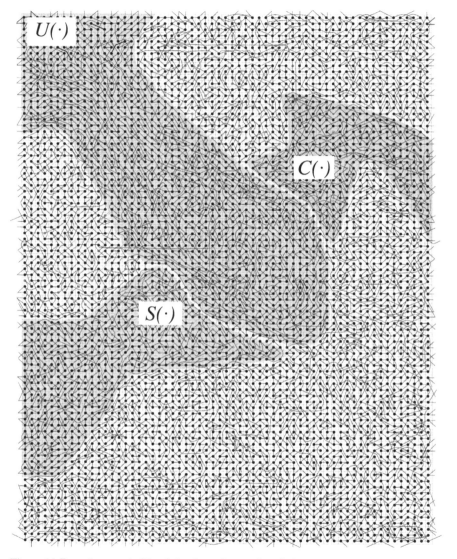

Figure 11.7.   A natural CA of the form 'network-cellular-automaton', including over 5000 structural primitives, each hooked to exactly three others. It is captured at a single computational 'instant'. It depicts a version of Figure 11.3 in a more abstract fashion at the deepest/finest granularity. Again, within (toroidal) universe $U(\cdot)$, scientist $S(\cdot)$ studies object $C(\cdot)$. P-consciousness is expected to be revealed in the exploration of the relationship between $S(\cdot)$ and $C(\cdot)$ in the CA context.

11.10 Formal systems, the natural CA and DAS

Formal systems are those that are constructed as a set of (meaningless) symbols that are established with a set of initial conditions (axioms), that can aggregate by rules of 'formation' (syntax), and that can merge, split, grow and shrink according to rules of 'transformation' (grammar). For a brief, simple introduction see 'Gödel's Proof' [E. Nagel *et al.*, 2002]. For a very floral, artistic and yet practical coverage, see 'Gödel, Escher, Bach : an eternal golden braid' [Hofstadter, 1980]. Remember, we have three things in dual-aspect science: (1) underlying reality $U(\cdot)$, (2) structure-aspect descriptions T', and (3), appearance-aspect descriptions *T*. In relation to each of these:

(1) If you haven't realised already, if ever there was a definition of a formal system, *the underlying universe $U(\cdot)$ is exactly that.* It's just that DAS says we'll never actually see it or exactly get it scientifically. We can get close.

(2) The structure-aspect is also a formal system. All the computational aspects covered here literally define a formal system. What is unclear is the extent to which it depicts $U(\cdot)$.

(3) As for the appearance-aspect? Whatever it is, it's not formal. The appearance-aspect is a non-unique, degenerate mapping of how $U(\cdot)$, appears. If it happens to 'appear' similar to a formal system, in a specific context ... so what? That does not make the appearance-aspect formal. It just makes an appearance-aspect scientific characterisation more abstract and perhaps useful.

A whole raft of previously endlessly debated hypotheses such as the Church-Turing Thesis, acquire new (and probably more decisive) levels of debate under DAS. For example the question 'can a computer be conscious?' is easily answered in the negative, because computers, as we currently construct them, completely disconnect the function of the computer from the vertical part of the structure-aspect hierarchy's chain of causality. It is that part of the chain of causality that makes us P-conscious, and it is gone in a computer. It is replaced by what we observe (appearance-aspect) as the electrical noise of the electronics (the actual physics of the computer. Whatever 'it is like' to be the computer, it has nothing to do with what we program into it. Claims that the remnant causality of a computer can approach human-level intelligence can now be seen to have a more formally visible Achilles-heel (lack of P-consciousness) not visible under the appearance-aspect. Can some artificial chunk of the actual natural world $U(\cdot)$, configured as per structure-aspect principles, be conscious? Of course it can. But that chunk of the natural world cannot be a computer of the traditional stored-program kind we presently use.

In addition, a computed 'appearance-aspect' model of the natural world has absolutely nothing to relate it to the natural world to which the appearance-aspect model refers. The physics of the original is missing! Along with it go all the original properties. Computational exploration of appearance-aspect set T members may be very informative, however, in its revealing of natural world appearances. That computation, however, is not an instance of the original natural process. The same goes for the computation that is the exploration of the structure-aspect statements t'_j. The structure-aspect computation revealing of the basis of intelligent (scientific) behaviour is not an instance of that behaviour, nor is it an instance of the original natural process expressing the behaviour.

DAS offers an almost surgical excision of the doubt that has fuelled the endless debates on these matters. Can you see the irony? Yes you can make a conscious machine, and an intelligent machine (one that can be a scientist). But it's not going to happen with a computer. It never could. To do it you must use the natural world itself, not a model of it (in either aspect of DAS). Whole books could be written on this one issue. I leave it up to them to elaborate it further.

11.11   The uniqueness of set T'

The (non-)uniqueness of set T has already been covered in Chapter 8. The question of uniqueness of set T' begs the question of structural primitives. In set T' construction, the structural primitives and intial conditions are uncertain and the set T' statements are computed and have the level of certainty of formal computation. Say 1000 scientists posit

1000 different structural primitives. They then construct 1000 individual sets T' which, it is claimed, result in a structure-aspect depiction of our natural world with a scientist/observer in it like us. Can this happen? It may prove to be that only certain structural primitives do anything interesting, and even fewer can be initialised to do anything as complex as an observer. There doesn't seem to be any obvious limit on the uniqueness of set T', but we won't know until we try. We may, in time, learn more and better ways to explore these alternate T' and increase the uniqueness of our T' science. It may turn out that after a huge amount of work, there are still 100 structural primitives and rule sets T', each of which is consistent with our set T and predictive of our P-consciousness. We can do no better. Fundamentally, however, we have a successful outcome: we have an explanation of some kind for P-consciousness and formal explanation has entered our science generally.

11.12 Alien structure-aspect science

We already know that appearance-aspect T_{alien} and T_{human} may be very different, depending on how different their P-consciousnesses are. But how does T'_{alien} compare to T'_{human}? There is one thing we know for sure: the universe, however it operates, literally produces both the alien scientist and the human scientist according to the same single set of structural primitives/rules. Does this mean that the structure/rule sets T'_{alien} and T'_{human} must come out identically? It seems that may well be the case. In a fully developed science, where most of the tractable mysteries were sorted out, it is hard to imagine how the alien's chosen structural primitives and T' could diverge from the human set and still be able to express both a human and an alien in a fashion consistent with their own particular P-consciousness and individual set T. It is expected that T'_{alien} and T'_{human} will converge once the T' set members are rendered invariant to the respective symbolic encodings of the alien and the human.

11.13 Qualification and implementation of DAS

I have no need to propose or prove my own or anybody else's choice of structural primitive, or set T' or explanation of P-consciousness. It is irrelevant to the task at hand. Any scientist can choose a structural primitive, do some computational exploration and make their neuroscience claim that establishes the validity of their set T'. The fact is that without the dual-aspect science framework all such claims are completely impotent. Only from the dual-aspect science perspective can the claims even begin to be discussed. The sensible appraisal of the dual-aspect science framework, via empirical testing managed by h_{DAS}, is a prerequisite to the dissemination of particulars of a proposed set T'. The non-uniqueness of T' tells us that there may be 100 physicists around the world, all of which might be able to construct their T' physics of some kind and make neuroscience predictions that explain the observer, and none of them will be heard unless dual-aspect science, or at least the process of examining its validity, provides the background framework.

11.14 Physical and material

DAS suggests interesting possible interpretations of words that appear often in the consciousness discourse. That which is 'physical' seems to refer to that which is in $U(\cdot)$ and described by set T'. This might act in contrast to the word 'material', which could be construed to refer only to that part of the physical world which has 'appearances' and is described by set T. Electromagnetic and gravitational fields are physical and not material. When something is 'non-physical' it might mean that it is entirely non-existent in $U(\cdot)$. Therefore consciousness and space are as 'physical' as anything else under DAS. This would imply that such views of P-consciousness as being non-physical or ineffable are misplaced. Food for thought.

11.15   DAS and the ultimate questions

All latitude to ascribe or impute analytical mathematics (e.g. as per appearance-aspect set T statements) as having any sort of structural role in the universe are gone under DAS. DAS explores underlying structure separately and thereby accesses causality. P-consciousness has an explanation of some kind. In that sense, the boundary of the knowable and known has increased. Yet despite this the underlying universe $U(\cdot)$ itself remains intrinsically unknowable other than to the extent delivered by the T and T' aspects. So in another sense the unknowability might be thought more pervasive because it applies to *everything*, not merely consciousness.

The limits of 'knowability' need a little more detail in respect of P-consciousness:

Q1 *"Why is to be human 'like something' at all?"*
Q2 *"Why do the experiences take on the particular qualities they do?"*

At this stage in proceedings, DAS only addresses Q1 by providing an empirically viable framework for propositions in answer to it. One or more fundamental principles may be expected. Question Q2 is left untouched at this stage. What can be said, however, is that Q2 will make no sense until Q1 has some kind of structure-aspect set T' presence. Note that to answer Q2 is not to deliver the actual qualities of the experience! That is a nonsensical expectation. However, when we have sufficient understanding of Q1 we may be able to sensibly state the circumstances behind a particular phenomenal 'feel'; the 'what it is like' of it.

One final prediction by DAS is worth mentioning. As physicists doing appearance-aspect science strip away more and more layers of organisation in a quest to get to more fundamental particles (= 'appearances'), the appearance is likely to begin to approach, at least in some kind of asymptotic sense, the operational form of the underlying structural primitive. In the Chapter 10 hierarchy analysis we saw the appearance-aspect hierarchy Population $\rightarrow$ Human $\rightarrow$ Brain $\rightarrow$ Cell $\rightarrow$ Molecule $\rightarrow$ Atom $\rightarrow$ Atomic particle $\rightarrow$ Subatomic particle $\rightarrow$ ... ... and so forth. The limit to the depth of the hierarchy is not defined. What the dual-aspect framework tells us, however, is that the appearance-aspect and the structure-aspect must converge in some way. The convergence does not invalidate the dual-aspect approach.

11.16   We're already doing it: The T' structure-aspect in existing literature

Some scientists have already done large CA explorations. James Crutchfield and colleagues explored simple 1-dimensional CA examples displaying complex persistent entities that were labelled 'particles' undergoing annihilative interactions $(x+y{\rightarrow}0)$, reactive interactions $(x+y{\rightarrow}z)$, spontaneous decay$(x{\rightarrow}y+z)$ and symmetry breaking [Crutchfield, 1994; Mitchell *et al.*, 1994]. CA are intrinsically expressive of easily recognisable physics processes. Crutchfield clearly but implicitly noted the difference between the T-aspect and the T'-aspect as follows: "*This spatial discrete hierarchy is expressed in terms of automata rather than grammars*". By 'grammars' he refers to appearance-aspect analytic forms. By 'spatial discrete hierarchy' he refers to the structure-aspect.

Stephen Wolfram published a monumental work on cellular automata called 'A New Kind of Science' [Wolfram, 2002]. Stephen Wolfram intuited the important distinction between the kind of science that a CA represents (set T') and the kind of science currently carried out by physics (set T). His book correctly identified the work as a 'new kind' of science, but then failed to pay attention to how to make it empirically stand on its own feet as science. There was no consideration of how cognition or the scientific observer might be explained or how science might be explained. All of it was offered as an alternate to what has been done already. There is no mention of subjective experience or the 'hard problem'. Nor was the question "*Under what circumstances might it be 'like something' to be an entity inside a cellular automaton?*" asked. Nevertheless, Wolfram's book is an excellent primer for a structure-aspect explorer.

In addition to Wolfram, all physicists that have worked on 'strings' e.g. [Sen, 1998], 'loops' e.g. [Rovelli, 2006], 'branes' e.g. [Ne'eman and

Eizenberg, 1995], 'dynamic hierarchies of structured noise' e.g. [Reginald T. Cahill, 2003, 2005; Reginald T. Cahill and Klinger, 1998; Reginald T Cahill and Klinger, 2000], 'quantum froth' e.g. [Swarup, 2006] and so forth finally have a potentially viable home in dual-aspect science under set T'. Also included in their group might be the 'artificial life' computational paradigm sometimes called A-Life. All that has to happen is that they re-structure their work into CA form by choosing a structural primitive, do the computation, and use the results to do neuroscience predictions involving the explanation of the scientific observer. So far they have not done this.

Consider the plight of physicists accidentally working in T'. It doesn't take much analysis to realise what happens when surfacing with ideas that match the existing empirical evidence and for which there is already a perfectly valid account in set T. They are perceived to have a more complex solution (based on a CA structural primitive) which makes the same predictions as set T. In a critical examination such proposals will be defeated with the empirical parsimony argument. The only way to win in this circumstance is for the new T' science to make predictions along the Figure 11.1 (h_{DAS}) route. Appearance-aspect science cannot do this, and it is this one place where the T' structure-aspect gains its status as empirical science. Having done that in neuroscience (in effect, predicting scientists), *then* their CA propositions can go head to head for compatibility with existing science elsewhere in the organisational hierarchy of the natural world.

11.17 Theories of everything (TOE)

The idea of a 'theory of everything' in appearance-aspect is fundamentally oxymoronic. As discussed here in detail, the dual-aspect framework predicts obvious impotence, of an appearance-aspect statement, in predicting a first person perspective (observer) or any explanation of how science is possible. This is rather ironic for a 'theory of everything' where 'everything' actually means everything except P-consciousness and scientists. So an appearance-aspect TOE is actually a misnomer. On the other hand, the structure-aspect is, by definition, a theory of everything. It does it all in one go (albeit perhaps computed for a very small subset of the universe).

11.18   Miscellaneous issues

Note that if the universe $U(\cdot)$ failed to produce a scientist, then no kind of science could ever occur. Universe $U(\cdot)$ may exist but will remain unwitnessed and scientifically unsung. $U(\cdot)$ may as well not exist. But if a scientist exists, that scientist cannot claim to have made $U(\cdot)$. Our presence inside $U(\cdot)$ clearly impacts $U(\cdot)$ and affects our observations of $U(\cdot)$. But to claim that observing $U(\cdot)$ *creates* $U(\cdot)$ is unsupportable under the only framework that allows the scientist inside $U(\cdot)$ to know $U(\cdot)$. Such considerations are interesting thought experiments that are technically moot and have no practical impact here. Believe it, don't believe it – it changes nothing.

The bottom line is that we live in a $U(\cdot)$ capable of creating P-conscious scientists. As luck would have it, at some point in our evolutionary history here on the boundary fringes of a dust mote in space called Earth, $U(\cdot)$ produced an entity with cognitive faculties sufficient to make science possible [Mithen, 2002]. Several million years later our behaviour was sophisticated enough to commence populating set T and embark on appearance-aspect science. This probably started somewhere between the time of the ancient Greeks and the renaissance. We have been doing appearance-aspect science ever since. Now, however, we also get to see that the way we have been operating is only a part of a bigger, more complete picture of the universe that includes how, and to what extent, it is possible to know your hosting universe from within it, when made of it.

Those that have read a lot of philosophy of science will note that, without regard, I have cut a swath through all the traditional ways of thinking about how scientific knowledge is acquired, what form it has, and what relationship the knowledge has with the actual natural world. DAS doesn't fit into any particular category. The entire thing is empirically constructed and can be tested/accepted/rejected empirically.

If I had to point a finger flaws in thinking, I would suggest a combination the following (not necessarily consistent) false assumptions:

(I) To assume that the underlying natural world is inaccessible and/or undescribable.

(II) To assume that the underlying natural world has no evidence basis.

(III) To assume that the underlying natural world is somehow rendered completely/literally visible/known.

(IV) To assume that what counts as scientific evidence is an identity with a report of the contents of an observer's consciousness.

(V) To assume that the scientific observer is unexplainable.

(VI) To assume that causality is unexplainable.

(VII) To assume that what is called induction is unexplainable (See Chapter 9).

(VIII) To assume that abstract order in language and/or logic and/or mathematics has any fundamental *determining* role in the natural world (as opposed to merely cohering with appearances when we massage it sufficiently).

(IX) To assume that natural order is/can be entirely captured by the communications tokens (like language) that we use to communicate.

(X) To assume that science accesses unique characterisations of the natural world.

The dual-aspect science proposition is what you have left when all these assumptions are gone. In terms of philosophy, I claim no expertise in knowing which philosophical category falls foul of which falsehood.

11.19 Summary

The large analysis presented in the previous chapters has culminated in a dual-aspect science (DAS) framework that is complex to operate by scientists, but simple to understand by those outside science. Under DAS, scientists make two different kinds of statements about the natural world. These are appearance-aspect and structure-aspect statements.

Appearance-aspect statements are those made by the bulk of science to date. The statements are sometimes called 'theories', and are expressed with various levels of mathematics or logic. Some are purely linguistic. Most are a mixture. Regardless, the natural world, when encountered by an observer, will appear consistent with the appearance-aspect statement, or the statement is somehow flawed. Appearance-aspect statements are therefore predictive to the kind of assumed observer that created them. Appearance-aspect statements encapsulate the apparent critical dependencies between objects but do not reveal why it is that things happen the way they do. Appearance-aspect statements are a product of the human mind and are expressed in forms created by the human mind.

In contrast, structure-aspect statements are not symbolic. They are not language. They are not mathematical symbols. They are not created by the human mind. They are the result a computation of a mathematical formalism designed and initialised by a human mind. They compute a description of the actual structural components of the underlying reality; a computation of the kind exemplified by a cellular automaton. That computation has the added burden of computing the observer (the mechanism of observation) as well as everything else. To isolate the operation of observation within the computed CA result, what needs to happen is to compare and contrast that part of the structure that is the observer's brain with that part of the natural world that it observes. In a comparison of one part of the computation with another we reveal whatever it is that is operating as P-consciousness. I make no assertions as to what this might be. I do not need to. The point is that unless you do the structure-aspect computation in the first place, P-consciousness will remain impenetrable.

The first thing that a structure-aspect computation proposal must do to gain acceptability is reveal a neuroscience mechanism (a prediction about the outward measurable appearance of brain material) that acts in explanation of a scientific observer that sees the world operate as per the appearance-aspect. This is a little self-referentially confusing, but is self-consistent and naturally complete (within understood limits) as system of knowledge.

Having done that neuroscience prediction, *then* the chosen computational structure-aspect formalism has the authority to make statements about other areas of the natural world, and each instance must also be consistent with the appearance-aspect. Examples of this are the traditionally understood appearances 'the atom' or 'space'. Structure-aspect statements are literally computer program outputs interpreted from the perspective of the dual-aspect framework, and they literally capture (in the form of another kind of description) the underlying causality of the natural world.

The structure-aspect explains why the world unfolds the way it does, and in particular how an observer built of it, within it, will describe it through an observation mechanism revealed within the computation. As a result, raw structure-aspect statements are likely to be quite difficult to encounter and interpret. Nevertheless, as we become adept at translating them into more recognisable appearance-aspect forms, the structure-aspect should normalise and cohere with the appearance-aspect, and become a routine part of scientific exploration.

We scientists are explained by DAS and, for the first time, scientists become the scientific evidence that makes the DAS framework possible. Letting ourselves become scientific evidence and letting ourselves be explained is the key to DAS. Because the scientific observer is explained, consciousness in general is also explainable under the DAS structure-aspect. In that explanation we will have sufficient information to make an artificial machine consciousness and prove it.

The DAS proposition is entirely empirical. It was designed to deal with an empirically measured deficit in scientific behaviour, and is proposed merely as a hypothesis h_{DAS}, ready for empirical testing as described elsewhere here. When the structure-aspect computation makes a prediction of the observable signs of the observation process, that are then measured by neuroscience, then the DAS framework earns its place.

> But not all theories are paradigm theories. Both during paradigm periods and during the crises that led to large-scale changes of paradigm, scientists usually develop many speculative and unarticulated theories that can themselves point the way to discovery. Often, however, that discovery is not quite the one anticipated by the speculative and tentative hypothesis. Only as experiment and tentative theory are together articulated to a match does the discovery emerge and the theory become a paradigm.
>
> [Kuhn and Hacking, 2012, Page 61]

Chapter 12

Scientifically Testing for Consciousness

Because it demands large-scale paradigm destruction, and major shifts in the problems and techniques of normal science, the emergence of new theories is generally preceded by a period of pronounced professional insecurity. As one might expect, that insecurity is generated by the persistent failure of the puzzles of normal science to come out they should. Failure of existing rules is the prelude to a search for new ones.

[Kuhn and Hacking, 2012, Page 68]

The following test regime is based on previous work [Hales, 2009b]. Here it is updated and calibrated with respect to the issues of this book. Otherwise, the basic procedure and underlying concepts are the same. The existence of the DAS framework immediately opens a door for a potential test for consciousness. The idea of scientific testing for consciousness is more than just a theoretical gambol. The inventor of a claimed artificial general intelligence is in need of just such a test regime, along with a principled relationship between P-consciousness and intelligence. In previous chapters I have explored the obvious logical consequences of a scientific account of the scientific observer, and a scientific account of how science is possible. I have reinforced repeatedly that when you use scientists as scientific evidence of our ability to do science – our P-consciousness is a critical dependency. Take it away and science stops. P-consciousness is, as I continue to stress, just a generalised form of 'observation' that is put to special use in scientific observation.

The obvious possible corollary is that scientific behaviour is proof of the existence of P-consciousness in the human scientist. Some logicians (likely not scientists) are probably having mild palpitations just now. Just because we humans use P-consciousness to enact scientific observation, and are critically dependent on it for scientific behaviour, does not logically entail that scientific behaviour can only result from having P-consciousness. This position is 100% true and simultaneously 100% useless to science. This is the 'I can imagine' argument. I can imagine scientific behaviour without the presence of P-consciousness. I can also imagine Higgs boson-related evidence without the actual presence of a Higgs boson.

The objective evidence for scientific behaviour is like particle accelerator evidence: the great buckets of all scientific statements (called T and T' in previous chapters) would have nothing in them if the P-consciousness of scientists was not involved. We have no actual evidence of truly original, authentic scientific behaviour arising without the involvement of human objectivity, which, as I have gone to great lengths to point out, objectifies-out-of subjectivity. The 'I can imagine' argument is therefore to be set aside in favour of using empirical evidence for argument. Perhaps if I qualify my claim a little: "Scientific evidence of *human* P-consciousness is demonstrated by scientific behaviour by *humans*". Maybe that is a little more acceptable. In any event, it is an empirical fact, so far, that human scientific behaviour critically depends on human P-consciousness. We may as well act as if that is the case, so that we may configure testing capable of proving us wrong.

What about testing something non-human, say agency X, for scientific behaviour? Would that prove that X has P-consciousness? It seems sound that if an authentic, original scientific act produced behaviour (a 'statement' of some kind) that a human might plausibly produce, then this supports a claim that X has at least the basic necessary P-consciousness that a human might have brought to bear in the same circumstances. This claim would be further supported by a clear case, by the human designers of X, of the relationship between X's ability to scientifically observe, the equivalence of the neurological basis of P-consciousness in humans, and the technological equivalent in X. It might also involve an argument that no other process of observation can be claimed to have occurred in X. This seems practical. So how do we construct a test in which no human judge is required, yet is an instance of 'scientific behaviour' that might plausibly be attributed to a human?

12.1 The paradigmatic doubt

The apparent inability to scientifically prove the presence, absence and kind of P-consciousness of any biological agent, inanimate objects, or technology such as a computer, has plagued the area of artificial intelligence ever since its inception. A valid formal scientific test does not exist. Let's start with a claim like *"proving consciousness is impossible"*. The claim can be rejected because there is no documented empirical or theoretical proof or principle underlying the claim. No citation of evidence is possible. There is only documented evidence of the claim of impossibility, which is equivalent to no scientific proof at all. It is merely baseless unscientific hearsay; a convention of the same kind as 'man cannot fly'. One is entitled to wonder what capacity of doubt in science is so powerful that it might completely deny even the possibility of such a thing? Could it possibly be a tacit presupposition about the nature/scope of science?

The usual rebuttal of claimed scientific proof of consciousness goes along the lines of *"I can't prove you are conscious and vice versa"*. It's easy to elicit in scientific circles. I have encountered it many times. When you investigate it, you find the position is based on a systemic lack of attention to scientists and how we go about our business. If we turn the proposition around, and accept scientists as scientific evidence of its necessary faculties, then if you hear *"I can't prove you are conscious and vice versa"* from a scientist, then the very evidence needed to do just that is telling you it is impossible find evidence! It seems rather ironic how invisible the evidence is when you are both a scientist *and* the evidence you seek. To deny scientists are P-conscious is to deny *'that which is scientifically seen'* the derivative status of *'scientific evidence of seeing'*. This is the faulty logic entailed by the (methodological) denial that scientists have P-consciousness whilst demanding it be used and being critically dependent on it in a verifiable fashion [Hales, 2009b]. If indeed this proves to be the case, then it is an almost textbook example of the most Kuhnian of Kuhnian paradigmatic blind-spots in scientific history. It can be immediately seen as a side-effect of the implicit training of scientists, by imitation, to be unable to include scientists in the natural world, and operate well *because* of that very circumstance.

Let's assume we designed a test that demanded scientific behaviour of a test subject. We all agreed that if the test subject did the science, then we claim it to be P-conscious. To see how culturally awkward a 'test for consciousness' feels, put yourself in the observer's shoes and ask *"Yes I agreed to the test procedure, but how can I be sure the test subject has visual P-consciousness?"* Agreeing to the test is one thing. Being asked to accept that what has been observed in the lab proves that P-consciousness exists within test subject is another. The main answer to this is that the scientist has to formally and finally accept that 'scientific observation' (of the literally eye-balling kind) and 'scientific evidence' parted company in science a long time ago. We have been accepting tenuous and complex chains of causality involved in things we have never actually seen to install statements in set *T* for many decades if not centuries. This one is only different in that it deals with the 1PP for the first time. Beyond that I can offer no solace.

This is a symptom of crisis-science and anomaly of the kind experienced by the first-entrants to a new paradigm. Perhaps we should feel privileged to be thus challenged. Those that doubt the test principles based on unjustified imagined alternative accounts of the evidence are advised simply to ensure the test is optimised to support P-consciousness claims with the same standards used elsewhere in science. One day we will be able to implant artificial P-consciousness hardware in brains. Then there will be a verification of the change to 1PP. Until then we must simply rely on logic and common sense.

12.2 Existing tests

The most famous of all such tests is actually not a test for consciousness at all. Alan Turing created an 'imitation game' in the 1950s [22, 23]. It is now called the 'Turing Test' (TT) and there remains an actual event called the Loebner Prize that carries out the test. A human judge interrogates via a terminal. Hidden, on the other end, there may be a human or a computer. Computer and human communications are then contrasted by the judging human. Nowadays we call this activity an interaction with a 'chatbot'. The structured mistake that makes the competing dialogues indistinguishable was supposed proof of equivalence of machine and human intelligence. It is not a test for consciousness, and makes no explicit link between consciousness and its role in intelligence. It's interesting and technically challenging, but here it is simply irrelevant.

In 1991 Stevan Harnad proposed a modified Turing Test called the 'Total Turing Test' (TTT), which requires that:

> "The candidate must be able to do, in the real world of objects and people, everything that real people can do, in a way that is indistinguishable (to a person) from the way real people do it."

Stevan Harnad [Harnad, 1991]

Clearly this is not a practical or a scientific test because it remains disconnected from a scientific principle behind consciousness and its relationship to behaviour (or the role of consciousness in intelligence) and it has no objective criterion for acceptance. From 2001 there is a 'Lovelace Test' (LT), which examined creativity/originality as an indicator of intelligence [Bringsjord *et al.*, 2001]. Again, like the (TT) and (TTT), in the (LT) a human judge determines the result, making this unscientific and basing the outcome on yet another presupposition about a relationship between creativity, intelligence and consciousness.

In 2003 Aleksander and Dunmall published a collection of five behavioural axioms and declared that if an artificial device design 'respected them', then the artefact is conscious. A test subject must incorporate evidence of

(1) *Depiction of sense of place*: The test subject has perceptual states that depict the test subject as embedded in an environment.
(2) *Imagination*: The test subject can recall previous sensory experience and can imagine potential experiences.
(3) *Attention*: The test subject can select sensory events of interest.
(4) *Volition/planning*: The test subject can imagine the results of taking actions and select an action based on goals.
(5) *Emotion*: The test subject uses emotional states as a means to evaluate actual or planned actions.

All of us might agree that these things say a lot about the properties of a conscious agency. However, the five axioms are a tautologous set of design requirements, not revealed causal relationships of the natural world that might create the characteristics of each item. Yet again, some kind of human adjudication determines the presence or absence of each crucial property [Aleksander and Dunmall, 2003].

If you look at the literature trail, what links all of the above propositions together is twofold: (i) A single non-scientific presupposition: that an artificial consciousness and/or an intelligent machine involves the use of computers and (ii) None of the propositions have any actual attachment to some kind of natural physical principle. In every case there is no science going on at all. There is merely engineering going on based on principles that have not been scientifically established.

In 2011 there was a sign of hope when Christof Koch and Guilio Tononi, both originating in neuroscience, published 'A Test for Consciousness' (KTT) that at least shows an appreciation of the actual problem, and identified something testable: *only conscious machines can be confused about unnatural incongruences.* The claim is that when a machine can detect 'that which does not belong', where the detection involves a physical impossibility (such as a man looking at his watch, floating hundreds of meters horizontally above a farm), then the only way that confusion can occur is if the machine has P-consciousness, through which expectations of the natural world have been established. Clearly that machine has to have a well developed sense of what is normal (in the natural world) so that the radically bizarre/abnormal can be a source of confusion. In principle I find this approach to be more in the spirit of my proposition here, albeit lacking in technical specificity that allows the test to be a 'consciousness detector'.

The Koch/Tononi proposition is grounded in the 'information integration' (an ABC-correlate of consciousness) theory of consciousness that was discussed in Chapter 6. The Tononi 'information integration' idea offers the prospect of a consciousness meter, in that 'integration', where it can be shown to correlate with P-consciousness, has a formal measure in entropy terms. Therein lies the prospect for some kind of measurement [Koch and Tononi, 2011]. In the KTT, therefore, there is an attempt to associate (i) some kind of measure of consciousness with (ii) an ability for a machine to be confused about an encounter with the natural world. While having already concluded that there can be no actual explanation of consciousness here, at least there is some kind of principle underlying a test that is also recognisable as science. It does have some prospect of being useful when a science of P-consciousness appears. However, for the purposes of formal scientific testing for machine consciousness, there is insufficient procedural detail and rigour.

12.3   The 'P-consciousness Test' (PCT)

The PCT can be viewed as an empirically viable variant of the TTT and LT obtained by choosing a single very specialised behaviour: *scientific behaviour*. I have already justified accepting scientific behaviour by X as conclusive proof of P-consciousness in X. A sound basis for testing is implicit in this. The idea is that the PCT itself cannot be passed unless the test subject has P-consciousness. I know there will be a huge debate on this position. In its practical implementation a test subject delivers scientific behaviour in a laboratory circumstance while we PCT architects remain assured that a test subject cannot behave scientifically without the requisite P-consciousness. To a certain extent the conflicts surrounding these issues is likely to only be resolved by doing the PCT as part of a more mature overarching strategy to deal scientifically with P-consciousness.

A machine that passes the PCT will naturally exhibit the properties of a conscious entity subjected to the KTT. Perhaps in the future this will be enough. Meanwhile, something like the PCT is to be preferred as an initiator of confidence levels. The benefit of this choice is that the PCT result is a signal from the test system itself, not a judgement by a human. This is no simple test, but it is in-principle practical and has logical teeth. Any embedded, embodied machine, no matter what the design basis, can be subjected to the test. In principle, animals and humans can also do the test, although it is procedurally more intricate. Ironically, the process of training mammals (e.g. a chimp) to participate in ABC-correlates of consciousness scientific experiments is actually a form of PCT. That means that the successful training of the chimp proves the chimp is P-conscious. It would be somewhat perverse to extensively train a chimp to use its P-consciousness in a scientific experiment directed at the study of P-consciousness, and then deny that the presence of P-consciousness is evidenced by the training process. The PCT is actually quite like the chimp-training process, but applied to a putative machine consciousness. The main difference is that chimp training is heavily supervised by humans. This is not allowed in the PCT.

The version of the consciousness test specified here proves the presence of visual P-consciousness. That is how the test subject is to acquire scientific evidence. To carry out a scientific act, it may be that the test subject has a multitude of other kinds of P-consciousness. However, successful test completion does not claim to have verified their presence. The successful test subject is only claimed to have delivered evidence of the visual P-consciousness that underpinned scientific evidence acquisition. That done, then other P-consciousness fields are validated as a side-effect.

The testers do not actually have to know anything at all about the physics that the creators of the test subject attribute to the origination of P-consciousness. The idea behind this is that anybody can construct a putative physics of P-consciousness. If they use the principles of that physics to construct an 'artificial scientist' of the kind required by the PCT, then their physics gets evaluated to the extent that the PCT test subject handles the test. Passing the test delivers evidence that the designer has found and adequately developed the use of the necessary physics of P-consciousness. Alternatively, failing the test invalidates the design and possibly the physics underlying that particular design. A human scientist becomes a familiar wild-type positive control or test benchmark. An artificial scientist would be the formal test subject. Test subjects must be exposed to circumstances neither testers nor tested have previously encountered. This gives the process the same level of double-blinding used in standard clinical scientific testing. What is required now is some more of the physical details.

12.4 Volition and the PCT

PCT test subject designers have an onerous job. In effect they must be deeply committed to the role of P-consciousness and the physics of its origins (or its lack thereof). Many problems have to be solved to make the agent sufficiently intelligent and mechanically dextrous enough to even enter the test. One of the biggest problems is the motivation to cause the PCT test subject to autonomously explore, then construct and deliver a 'law of nature'. Without motivation, the PCT test subject will fail. By what means is the test subject motivated? Scientific behaviour is complex and results from abstract goals. In humans these goals are optionally, indirectly related to basic physiological necessities. Complex motivation demands an onerous but in principle practical level of development of the faculties of the test subject.

Normally, to motivate an animal to behave in a certain way uses a reward/penalty mechanism. The PCT test subject can be similarly motivated by the act of scientific behaviour being linked to a penalty/reward result. What will probably happen is that the subjective qualities of hunger, breathlessness, thirst, orgasm, pain and others may be engineered and used to motivate the test subject. Some kind of experience might be attached to the power supply energy level. Combined with ever-decreasing performance capacities, experiences like hunger or pleasure reinforce behaviour that involves foraging for energy that in turn demands, implicitly, a scientific act. It sounds ethically problematic, but until the test is successfully completed we are in no position to argue the ethics of it. Indeed in a perverse way we are ethically bound to do the exploration so that future scientists will have clarity in the matter. One way or another, the problem of volition and related behaviours is an onerous and central problem that must be solved for the test regime to operate properly.

12.5 Embodiment and physical implementation

It will shortly be shown how, inside the test rig, radical novelty will be required. Its implementation will involve 'laws of nature' different to that of our day to day lives in ways the test subject designers cannot be allowed to know. Test survival is likely to require test subjects meet certain minimum electrical, mechanical and environmental standards. That is all that the test subject designers can be told. Nor can the test subject designers be told about the actual environments to be experienced by the test subject. There will be limits on such things as volume, weight, centre of gravity, centre of mass and other properties when travelling and when stationary, and so forth.

There will be a minimum requirement that the test subject be able to move at a certain speed within a certain range in various terrains inclusive of certain topological features. It may be required to right itself from certain orientations or operate in no gravity or altered gravity. There may be a specification on grip strength and an ability to carry an object or ranges of objects of a certain class (size and weight) within a certain reach. There may be a minimum set of motions like 'jump'. Perhaps specific sounds or other indications may be required. Basic senses may specify an ability to see/hear within/outside human frequency ranges. The test subject may be required to collect and carry/deposit materials of certain kinds and so forth. These details can be set aside as they are a design issue for the future.

PCT developers must design multiple test environments. The role of each will suit the specific needs of any individual stage of any test trial. The manner of ingress/egress for each environment shall also be built into the PCT test rig. The exact nature of the test environments is otherwise open. It may include such facilities as a maze or obstacle course with/without mirrors. The test subject may be a spectator of some sort of activity and/or a participant in some sort of activity such as a game. It may be possible that humans could, with suitable preparation, operate as wild-type control subjects by encasing the human in hardware limiting all behaviours to that equivalent to PCT standards. This can be used to construct test execution benchmarks. The whole test may have a time limit set by human PCT testing.

12.6 PCT overall strategy

It is scientific behaviour that is the sought-after key to P-consciousness claims, and we know that this is no simple thing. How do we cope with getting a PCT test subject to come up with an original scientific act, followed by delivery of an appearance-aspect statement of kind t_n for depositing in set T? It sounds appallingly complex. Does our machine have to win a Nobel Prize? The answer to this, after a little reflection, is definitely *no*. The complexity of the science does not matter! Simple science or complex science is still science that is just as critically dependent on P-consciousness. In Chapter 9 we covered the dynamics of belief formation as involving the adaptation of a state-trajectory expressing the belief. The basic belief was written as

$$\textbf{Belief}(t) \tag{12.1}$$

A change in the state-trajectory-expressing belief is represented by

$$\frac{d\textbf{Belief}(t)}{dt} \tag{12.2}$$

The successful completion of the PCT need not require the subject to deliver an absolute 'law of nature' (a belief captured as equation (12.1)). Rather, we can look at the before-after change as per equation (12.2) for a simple 'law of nature'. To do this, we first check for the non-existence of pre-existing belief (i.e. ignorance of a target belief). We then give the test subject scientific evidence of the potential belief. In effect, we are training the scientist in a 'law of nature'. As a result we have used equation (12.2) to eliminate the need for some really complicated belief. The test subject's internal set T has, as a result of being in the test regime, been subject to a verified change from ignorance to knowledge.

The next difficulty is delivery of the abstract form of the 'law of nature'. On reflection, the exact abstract form of T that a human might deliver is also unnecessary. Remember, a single human can do science. In the absence of anybody to receive the new statement, the existence of a formally externalised abstract representation of the novel 'law of nature' can be demonstrated by encoding it in the more complex form of the test subjects' solution to a puzzle that cannot be solved unless the correct abstract form of the 'law of nature' was acquired. We can get the test subject to do a 'law of nature dance' that can only be possible had the abstract law of nature been properly acquired in the first place. That dance might not be a statement of the kind publishable in a scientific journal, but it certainly communicates the scientific knowledge that might have been acquired by someone reading the scientific journal. What we have done is exploit our understanding of the dynamics of belief change in a P-conscious agent.

This process seems quite practical. We cannot make it too simple, however. Indeed it is a requirement that the law of nature expose the test subject to radical novelty. It cannot possibly have been trained in the results before attending the test. Nor can the tester know what it is. The completion of the test must be carried out by the test subject entirely under its own volition. We now have the basic design of the test:

(1) Introduction into the test environment.
(2) Verification of ignorance of the target scientific statement.
(3) Exposure to a radically novel 'natural' regularity that the test subject is to acquire.
(4) Application of the item (3) 'training' in a completely novel circumstance.

On the theme of belief dynamics, we can also demand that multiple PCT trials, repeated in various orders and in different novel circumstances, shall reveal no drift or interference with prior learning. That is, groundedness in P-consciousness shall prevail in the Belief(t) system dynamics of successful test subjects to the same extent that it does in human control subjects. Problematic learning dynamics should reveal itself in convergence and stability errors. The use of problem solving to verify correct learning takes advantage of the generalised 'statement' as being any behaviour by the test subject. The need for *a priori* installation of communications such as language is thereby eliminated.

12.7 'Radical novelty' and its forms

I have illustrated the most basic PCT test procedure, and I will go through a worked example shortly. Before I do that it is worthwhile to look at some special examples of 'radical novelty' and how they might be brought to bear on more sophisticated PCT learning and problem solving scenarios [Hales, 2009b].

12.7.1 *Communication as problem solving*

The literal creation of an ability to communicate with another agent or human scientist is identical solving a novel problem. This is suggestive of a further class of testing that could form part of an overall test regime: the ability to actually *create* a novel communications protocol, not merely demonstrate or use an existing one. Twin PCT test subjects could be required to devise and then communicate novel knowledge to each other under the basic PCT defined above. This is not a new idea. A number of existing AI test regimes already use prototypical versions of this process [Gamez, 2008], although their success/failure are not claimed as conclusively revealing of P-consciousness. The PCT adds behavioural demands that make the communication behaviour evidence of P-consciousness.

12.7.2 *Mirroring as problem solving*

Another suggested aspect of a PCT is a communication behaviour called 'mirroring', where experience of the distal external world is demonstrated through literal mimicry. In an encounter with novelty, mirroring is also special communication with a dependence on self image. Construction of self-knowledge intrinsically demands an externalised P-consciousness of the kind discussed. In the construction of an externalised projection into the distal world, one of the objects in that projection is the external portions of the test subject itself. The agent constructs self-knowledge the same way it constructs knowledge of anything else. Self-knowledge revealed through mirroring may thereby form another potential route to a demonstration of P-consciousness.

12.7.3 *Mirrors and problem solving*

Mirrors are a way of directly introducing and controlling perceptual errors of the kind referred to in the KTT. Encountering and successful problem solving in respect of a mirror is critically dependent on an externalised P-consciousness depiction of where the mirror is located so that self and NOT-self can be properly distinguished. The test subject cannot do this without an internal phenomenal representation of the external world enabling the mirror itself not to be confused with the image it contains. Just like in the KTT and somewhat paradoxically, to literally be mistaken in certain predictable ways about a mirror might lead to a short-cut way to prove the existence of a P-conscious representation of a mirror.

12.7.4 *Hardware intervention*

Electronic knowledge erasure and exogenous control of the P-consciousness (enable/disable/alter) of the PCT test subject could be used to effect repeat trials and recreate radical novelty in a variety of ways.

12.8 PCT overall execution logistics- single trial

The following procedure can be found in its original form in [Hales, 2009b]. It covers a single trial of the basic belief dynamics for a single test subject $S(\cdot)$. Clearly there is scope for great elegance in future test scenarios and methods. For illustrative purposes, and to make the point of this chapter, all that is needed is the barest of minimal PCT rigs. It doesn't use the communications, mirroring, mirrors or hardware intervention options. The basic flow of the simplest single PCT trial will involve probably three stages and use two environments as shown in Figure 12.1. The two environments are each imagined populated with a small swarm of remotely operated or autonomous vehicles. They are 'dressed' to suit the occasion and to present various levels of difference/similarity consistent with the severity of the challenge. These are used to cue/train the test subject.

There must be a reward mechanism that the test subject is already aware of because it is part of its basic operation. This may be an ability to recognise and connect to a familiar adaptor, from which energy derived. This is shown in Figure 12.1.

12.8.1 *Stage 1 – The reward room*

Stage 1 introduces test subject $S(\cdot)$ to one particular environment $E_i(\cdot)$ never before encountered by $S(\cdot)$. Embedded in $E_i(\cdot)$ are two things (a) a reward system and (b) a system of perceptual cues in the environment that encode an abstract statement t_i. Statement t_i will later be discovered by $S(\cdot)$. The intent is that $S(\cdot)$ recognise and acquire reward through demonstration of the abstract science knowledge t_i. At this stage the reward mechanism cannot be recognised or attended to by $S(\cdot)$ because it will never have been exposed to it before and cannot be recognised. The environment $E_i(\cdot)$ and statement t_i will be sufficiently complex that random exploration behaviours by $S(\cdot)$ are extremely unlikely to result in a reward within the allocated duration of this test phase, which will be calibrated based on human trials. As such, $S(\cdot)$ will fail to acquire the energy reward, and will expend non-trivial amounts of energy reserves in the process. However, the belief dynamics of $S(\cdot)$ will result in familiarity with $E_i(\cdot)$, which will implicitly result in knowledge of signs of the reward mechanism located therein. At some point in the process, triggered by some set of events to be decided, access to a second novel environment $E_j(\cdot)$ will become available and $S(\cdot)$ will egress to it and be captured by it, unable to return to $E_i(\cdot)$.

12.8.2 *Stage 2 – The science room*

Environment $E_j(\cdot)$ repeatedly and automatically demonstrates regular behaviour according to the abstract statement t_i. This behaviour is completely equivalent to a natural 'law' in all respects except that it has been created by humans. The 'natural world' that is $E_j(\cdot)$ operates according to t_i in the same way that Newtonian dynamics operate, for example. The key attribute of the science room is that when the sequence of events encoding t_i occur, $S(\cdot)$ will literally be rewarded, exposing $S(\cdot)$ to those aspects of $E_j(\cdot)$ that are responsible for reward. In being exposed to the science room behaviour, $S(\cdot)$ will be exposed to t_i via specific causal relations located distally from $S(\cdot)$'s boundary. In Figure 12.1 the objects assemble themselves in the 'Knight's Waltz' configuration from chess: <1 step, RIGHT-90, 2 steps>. The reward ensues and then the objects disperse. Some time later the process repeats.

Many 'laws of nature' of the general kind '*only after this event*', '*number N of these*', '*only when that object and that object together do this*' or '*when this pattern happens then*' and so forth, are possible. The successful learning dynamics of $S(\cdot)$ will result in the encoding of an implicit abstraction equivalent to knowledge of t_i and its association with signs of the causal antecedents of reward. The process of inhabiting $E_j(\cdot)$ will also cause non-trivial amounts of energy expenditure. The reward process will be superficial; insufficient to recoup losses. At some point in the process, triggered by some set of events to be decided, the activity resulting in rewards will cease and access to the reward room environment $E_i(\cdot)$ will become available. $S(\cdot)$ shall return to $E_i(\cdot)$, hopefully with brain dynamics reconfigured with the newly acquired knowledge of the outward signs of the reward system.

12.8.3 *Stage 3 – Return to the reward room*

Having returned to $E_i(\cdot)$ the test subject now recognises the mechanism of reward. Mere demonstration of a desire for reward will not be sufficient demonstration of P-consciousness-mediated learning. For example the agent $S(\cdot)$ may adopt the physical behaviour involved during the act of receipt of reward, such as physically connecting itself to some feature inside $E_i(\cdot)$. The reward shall not be forthcoming unless the abstract rule t_i is used in a completely novel context. In Figure 12.1, $S(\cdot)$ is expected to reproduce the Knight's Waltz with completely different objects. In the reward room $E_i(\cdot)$ will be signs that the abstraction t_i will result in reward. Behaviour by $S(\cdot)$ that demonstrates abstraction t_i will activate the reward system. Only the experience of the science room distal $E_j(.)$ can supply the key concept. The test subject will be given a time limit, again calibrated by human trials, to activate the reward system. The activation of the reward system or the expiry of the time limit terminates the test.

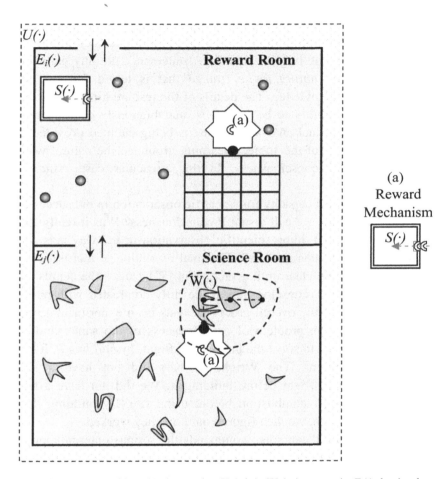

Figure 12.1.   Test subject $S(\cdot)$ learns the 'Knight's Waltz' pattern in $E_j(\cdot)$, having been already checked for ignorance in $E_i(\cdot)$. The energy delivery system is hidden inside the polygon and is accessible to $S(\cdot)$ when the pattern shown as $W(\cdot)$ is assembled in contact with the black dot on the polygon. The $E_j(\cdot)$ objects randomly and repeatedly assemble themselves in various configurations of the knights waltz moves. The 'law of nature' t_i is the underlying pattern. Assembly of the objects in the pattern appropriately aligned with the marked polygon vertex results in reward. Based on [Hales, 2009b] Figure 3.

12.9 S(·) passes the PCT. What next?

Because of the nature of the test, and, it is claimed here, because of the role of P-consciousness in the kind of learning that is scientific behaviour, a test subject cannot be trained to do any particular PCT. A subject dependent on prior training will fail. The test subject is required to deliver an authentic original scientific 'statement', not any particular statement. It is *an ability to be trained* that is tested for, not any particular skill or knowledge. The details of the test are highly contrived to ensure that the ability to be trained is unambiguously demonstrated. That ability, in the final analysis, is what is being claimed provided by the P-consciousness of the subject: learning grounded the natural world as presented by P-consciousness. In this particular case *visual* P-consciousness.

But is the claimed capacity for scientific observation in the successful $S(\cdot)$ identical to what we call visual P-consciousness? Was it really 'like something' to be $S(\cdot)$ doing scientific observation in the way it is 'like something' to be human scientist? The final resolution of that issue will probably not become clear until a successful PCT causes the details of a putative physics of P-consciousness to be fully contrasted with human brain physics. What the overall process reveals is an experimental route to the solution to the problem of consciousness of the same kind the Wright brothers used to solve the problem of flight: *by building it*. It is in principle no different. The Wright brothers did not have a fully developed theory of flight before building it. We did not have a fully developed science of combustion before using fire. By building these things and using them, we *then* figured out how they worked.

A successful PCT candidate commands the serious entertainment of the idea that P-consciousness was essential in its success. Prior to this, all we had was conjecture, presupposition and no scientific way forward because science itself wasn't configured correctly. Ultimately it is the reconfiguration of science to dual-aspect that leads to insights about us, scientists, that makes sense of testing for P-consciousness.

Having got to this position it is sobering to step back and look what the test is doing: nothing more than very basic puzzle solving intelligence tests of the 'normal science' kind that Kuhn cites all through his *Structure*! This time, however, it was carried out by a *machine*. To make this happen our perspective and understanding, of how we humans do puzzle solving 'normal' science, had to change. In that change, something that would otherwise seem unremarkable in day-to-day life looked different; had different implications. As a result of the change in 'world-view' we have been able to configure a test in a special way that makes a previously incorrigible natural world behaviour give up its underlying regularity. This is precisely the effect cited in the many Thomas Kuhn quotations included in this book. The same world now looks different because we have the perspective of a different paradigm; a way of thinking that now includes an understanding of how scientists do science.

> Discovery commences with the awareness of anomaly, i.e., with the recognition that nature has somehow violated the paradigm-induced expectations that govern normal science. It then continues with a more or less extended exploration of the area of anomaly. And it closes only when the paradigm theory has been adjusted so that the anomalous has become the expected. Assimilating a new sort of fact demands a more than additive adjustment of theory, and until that adjustment is completed – until the scientist has learned to see nature in a different way – the new fact is not quite a scientific fact at all.
>
> [Kuhn and Hacking, 2012, Page 53]

Chapter 13

The Kuhnian Take: Wrapping Up

But for normal scientific work, for puzzle-solving within the tradition that the textbooks define, the scientist is almost perfectly equipped. Furthermore, he is well equipped for another task as well – the generation through normal science of significant crises. When they arise, the scientist is not, of course, equally well prepared. Even though prolonged crises are probably reflected in less rigid educational practice, scientific training is not well designed to produce the man who will easily discover a fresh approach. But so long as somebody appears with a new candidate for paradigm – usually a young man or one new to the field – the loss due to rigidity accrues only to the individual. Given a generation in which to effect the change, individual rigidity is compatible with a community that can switch paradigm to paradigm when the occasion demands. Particularly, it is compatible when that very rigidity provides the community with a sensitive indicator that something has gone wrong.

[Kuhn and Hacking, 2012, Page 165]

Here I look at the big picture of the story for science apparent in the DAS upgrade, but from the perspective of trying to understand if we are indeed at the cusp, in the midst of a crisis state in science. I use the Kuhnian criteria and perspective applied to the circumstances revealed in the previous chapters. I also preempt the critique of the DAS proposition in the manner of Kuhn's response to critique found in the postscript of the 1969 edition of his *'Structure'*. Those that want to 'cut to the technical chase' can probably skip to the summary.

13.1 Anomaly and the signs of science revolution

The Kuhnian sense of the imminence of scientific change manifests in the appearance of anomaly. These are usually cast as scientific phenomena that cannot be explained by the universally accepted paradigm of the group of scientists to whom the anomaly appears. This sounds perfectly in accordance with the evidence that history provides down the centuries. This chapter seeks to more specifically identify the presence of anomaly in the current scientific era, and then justify the need for change.

The difficulty that the process encounters is that the particular anomaly in question is part of a paradigm that we scientists formally *do not even know we are in*. None of us explicitly chose to be in it. We inherited it implicitly by imitation of mentors, a scientific training process that installs tacit knowledge supporting accepted behaviours (Chapter 7). If there is no explicit recognition of the normal science of the paradigm (because it's not written down anywhere), then how is any scientist able to recognise anomaly? The paradigm being presented with anomaly is one that probably commenced with the ancient Greeks, but frankly, it doesn't matter exactly when it might have been established. Its establishment is long out of living memory, and that's all that matters for an undocumented, implicitly trained aspect of being a scientist to be significant in the eventual appearance of anomaly.

This anomaly is different in the sense that it applies to *all* scientists, not just those in the grip of a specialised science sub-discipline or even an entire branch of science like biology or physics. The anomalous paradigm is universal and implicit. It has already been revealed multiple times in this book. Science universally presupposes the observer within an implicitly trained behaviour that has led to all the scientific successes to date. I took the trouble to measure the successful behaviour and document it. This is the 'law of scientific behaviour' (the now familiar $t_A/t_n/T/H$ framework), in which the critical role of the observer is obvious and explicit. The story it tells is stark. Science has only ever described *what* (appearances/critical dependency). It has never explained *why* (structure/causal necessity).

I then looked at the obvious prediction, fundamental to the t_A behaviour: that science will fail, permanently, to explain the scientific observer. This is a double-barrelled failure. Science presupposes the scientific observer *and* has compounded that presupposition by making sure, systemically, accidentally, that the presupposition is permanently in place and never actually examined by scientists, who remain unaware of it.

Think about that outcome in the sense of the appearance of anomaly. This simply does not look like the sudden appearance of anomaly. If that failure is an anomaly in the 'Kuhnian paradigm' sense, then it has actually been there all along. It's not new. In its ability to account for a scientific observer, the paradigm itself was *defined* with anomaly in it from the start. What is new is the recent circumstance in neuroscience where the paradigm has begun to directly address the one and only intrinsically, paradigmatically impossible scientific account: the scientific account of the scientific observer of the kind needed for an engineer to build a scientific observer. It does so in complete ignorance that it will fail.

As an encounter with anomaly, this one is therefore, in a sense, backwards. Instead of discovering a novel aspect of the natural world that does not fit the paradigm, we have accidentally tripped over an anomaly that has always been there, and discovered we are in a paradigm we didn't know about. Usually the appearance of anomaly is called some kind of discovery. But this is not the usual discovery about the natural world 'out there' beyond the bounds of the scientist. This time, the 'discovery' is about *ourselves*. It inverts the normal process, but in some general sense it is just a matter of perspective. Someone may be happy to classify it as the same kind of appearance of anomaly. I'd rather get on with fixing it.

Having explicitly measured the paradigmatic scientific behaviour, and recognised its limitations, I then looked at what changes (a form of self-governance previously absent from science) might be made to *science itself* that might make accounting for the scientific observer a scientifically tractable activity. That change was not difficult to find, and in its creation and easily implementable exploration with computers, it introduced a dramatic hole in existing science. Fifty percent of the possible scientific behaviours (kinds of statements) are simply missing from science, and within the missing half is what is in store for us as a real science of consciousness, along with a second, validating view of every other law of nature we have ever established – one that includes an account of causality. The new view does not supplant the old, it sits alongside it.

One of the great arguments in the wake of Kuhn's *Structure* was the idea of incommensurability of the new paradigm with the old. The old paradigm could not predict the new and could not predict what the new paradigm predicts: that is usually where anomaly arises in the first place. Once the new paradigm is in place, sometimes there is a sense in which the old can, within certain limits, be couched as a subset of the new. Depending on the extent of the conceptual basis underlying the change, and how far-reaching it is, the new and the old are a bit like apples and oranges. The comparison does not make sense. Where does 'commensurability' fit into the proposed change to science, with its additional (second) realm of statement production?

The new framework, dual-aspect, adds a structure-aspect to the existing appearance-aspect framework. The new sits nicely alongside and does not invalidate the existing framework. It is obvious that the incommensurability of the new and old is actually *built in to the change itself*. Far from creating incommensurability and causing the discarding of the old, the new framework has incommensurability built into its operational fabric. There are two views of the same underlying reality, and their incommensurability is intrinsic to the co-existence of the two views. The two views must be consistent with each other. That mutual predictive consistency is demanded by the DAS framework as an operational requirement, and adds to the robustness and surety of all science outcomes, new and old.

From another viewpoint, the new framework can actually provide an explanation of the incommensurability experienced under the current single-aspect framework. Incommensurability is actually *expected* because the appearance-aspect simply organises appearances of the natural world. Those appearances have no content that spans any traditional intra-disciplinary paradigm shift, and also have no content that accounts for the 'operational paradigm shift' that is the vertical leap across the disciplinary boundaries (say from physics to biology and the apparently emergent properties that reveal themselves in that act). The disciplinary boundaries are simply absent in the new aspect of the dual-aspect framework. They have no meaning. The statements in set T' cannot stop expression at some point in the natural hierarchy like they do in set *T*. The structure-aspect set *T′* statements express it all or they express none.

The anomaly that I claim exists is therefore unlike any previous anomaly and presents itself as a suite of correlated and relatively diffuse symptoms:

(i)   The systemic lack of explicit recognition or even basic awareness that mentor/novice imitation acts at the heart of current scientific behaviour as evidenced in the (now) proved unexamined/undocumented state of the appearance-aspect science framework (which has actually been visible in the literature since 2009 [Hales, 2009a]).

(ii) The appearance-aspect framework results in scientific behaviour that presupposes the observer, making an account of the observer logically impossible *a priori*.

(iii) The use, for roughly 20 years, of the (i) framework by neuroscience in the 'ABC-correlates of P-consciousness' paradigm that is tacitly assumed to be the only kind of account of consciousness –and– without realising it is actually explaining scientific observation, thereby contradicting (ii).

(iv) The tacit reliance on a scientifically unjustified principle, the 'Mind Brain Identity Theorem', that simply asserts the impossibility of a separate explanation of the mind while recognising that the mind exists, and being 100% dependent on it for scientific behaviour in a verifiable way.

(v)   The waning presence of philosophy as the ABC-correlate science of consciousness grows (as predictably ill-fated as it is – Chapter 3). This is, according to the evidence of history, the typical state of paradigm-change-related science as shown at the beginning of Chapter 4.

These are posed as constituting a classic example of anomaly described by Kuhn as a 'special occasion prepared by the advance of normal research that finds it can only be conducted in an abnormal way'. The difference this time is that the abnormality was revealed by finding ourselves in a paradigm we were unaware we were in.

There are other more subtle signs of a crisis state. Consider the difference between the two statements: (a) *"Science explains the natural world using a dual-aspect framework"* and (b) *"Science creates descriptive statements using an appearance-aspect framework, and will never explain the observer"*. In one of these, science is a strange sort of club with a defined taboo central to its activities, and in the other science is fully engaged with a fundamental principle that recognises the limits to scientific knowledge, and at the same time has no limits on the choice of scientific target.

I am now a scientist. I now realise that when I became one I was enrolled in (b) without my knowledge, and remain there by accident of history, and have never been given the chance to argue for or against (a) (until now). Consider your status as a scientist. Is this an issue that might have used a little more attention in your science training? How do you feel about being tacitly enrolled, in effect, in a taboo-centric club that is unaware that it is one?

Another Kuhnian sign is the pre-paradigmatic 'science-of-consciousness' book-blizzard that has occurred since the inception of the 'ABC-correlates of consciousness' paradigm. This book is a snow flake in that blizzard! The dialogue in these books is a fine example of the continual reiteration of foundational matters in a potential science of consciousness. These books address other writers as much as nature, just like Kuhn tells us of pre-paradigmatic periods in the past. If the science upgrade to dual-aspect occurs, then it is likely that the appearance of the book blizzard will be re-interpreted as part of the change process for science as a whole, not just the science of consciousness. It acts as evidence of both.

A minor sign is me. Kuhn advises from the history of paradigm changes that the human agent of change is usually either young or new to the field [Kuhn and Hacking, 2012, Page 90]. Being from parts unknown to science (real-world commerce and engineering), I fit the profile. My status is no proof that a shift is afoot or right, but if the proposed shift happens, I'm sure future historians will enjoy the irony of it.

Another Kuhnian anomaly is the obvious presence of a whole set of physics propositions that fit the structure-aspect, yet are presented by physicists that don't know it. They fail to impress the existing appearance-aspect scientists (that don't know they are appearance-aspect) for all the reasons in the dual-aspect science chapter. They don't realise that their ideas are incommensurable (in the sense discussed above) with existing science for perfectly understandable, predictable reasons. They also don't know that their empirical justification involves making neuroscience predictions consistent with some kind of explanation of the scientific observer ... way off in neuroscience where most physicists (especially cosmologists) tread not.

> No process yet disclosed by the historical study of scientific development at all resembles the methodological stereotype of falsification by direct comparison with nature. That remark does not mean that scientists do not reject scientific theories, or that experience and experiment are not essential to the process in which they do so. But it does mean – what will ultimately be a central point – that the act of judgement that leads scientists to reject a previously accepted theory is always based on more than a comparison of that theory with the world. The decision to reject one paradigm is always simultaneously the decision to accept another, and the judgement leading to that decision involves the comparison with nature and with each other.

> [Kuhn and Hacking, 2012, Page 78]

Consistent with Kuhn's view, I have provided an alternate paradigm, dual-aspect science, and shown it to be practical in ways that were previously impossible because it involves the use of massive computational resources unimaginable until quite recently – well after the time of Kuhn, and which are now within the routine reach of most scientists. The full empirical validation of the DAS proposition also requires a level of sophistication in neuroscience that was not in existence in Kuhn's time. Structure-aspect science is purely computational, fundamentally different, and practical. It is a viable alternative that some scientists already use but don't know it. I posit this state of affairs as sufficient grounds that a shift to dual-aspect science, if in the offing, cannot be rejected for lack of being a viable alternative. This is especially so since it does not reject even one of the existing appearance-aspect laws of nature and has already been shown able to be established empirically and therefore to look like single-aspect science prior to the shift.

Yet more possible Kuhnian signs are (1) the continued failure to unify relativity (gravity) and quantum mechanics and (2) the existence of the patches evident in the 'dark matter' and 'dark energy' propositions. To the extent that these are compensating terms bolted on to existing equations to make sense of observations, they may yet be seen as consistent with Kuhnian paradigm-shift evidence.

These patches exemplify the process that Kuhn reveals in, amongst others, the Ptolemaic ⇔ Copernican revolution. By the time of Copernicus, the Ptolemaic system of celestial order had engaged multiple lifetimes of wrestling with residual discrepancy and multiple patches, to the point of its adherents being unable even to agree on a calendar year [Kuhn and Hacking, 2012, Page 83]. The Ptolemaic system was therefore critically vulnerable to anyone that saw the actual nature of the underlying order, and fixed it. Perhaps the modern physics patches are representative of exactly the same process. I can claim that maybe the dual-aspect approach is the hidden order that makes sense of the patches and incommensurabilities. Maybe quantum mechanics and relativity are destined to be incompatible in principle, merely because they are the products of appearance-aspect science. Their merging in the dual-aspect view of nature makes their incompatibility an expectation, and eliminates the whole conflict. Isn't it worth considering? Is it really worth ignoring without examination?

> Confronted with anomaly or with crisis, scientists take a different attitude toward existing paradigms, and the nature of their research changes accordingly. The proliferation of competing articulations, the willingness to try anything, the expression of explicit discontent, the recourse to philosophy and to debate over fundamentals, all these are symptoms of a transition from normal to extraordinary research. It is upon their existence more than upon that of revolutions that the notion of normal science depends.
>
> [Kuhn and Hacking, 2012, Page 91]

13.2   The DAS aftermath: a preemptive postscript

Thomas Kuhn was prescient. Ahead of his time, he saw the crucial centrality and primacy of the human brain and its perceptions in scientific behaviour and its products, enacted both as individuals and as a collection (a community) unified by a common purpose. We can now voice the reality of it because we have done the neuroscience. We know that what he intuited is accurate. As detailed by dynamical systems analysis, the fundamental transactional unit of the entire scientific edifice is the localised coherence of one cohort of neurons with another, within the brain state trajectory of one individual. That event is witnessed in a scientific statement by that individual. Mediated by the statement and its critique, neural coherences propagate throughout the entire paradigm-centric community. All the evidence points to this simple reality. The detailed operation of it has been covered in previous chapters on dynamical systems, consciousness (observation), hierarchy and cultural learning theory. It's all there in detail not available to Kuhn, yet he understood and/or intuited the presence of all of it and its impact on science, and when he saw history laid out before him, that was the story it told him.

That said, the way paradigms were viewed after Kuhn's *Structure* has been largely re-engineered by subsequent reviews, including some by Kuhn. Post 1962, there was a large discourse (it still continues! E.g. [Weinert, 2013]) that disagreed that the evidence Kuhn presented supported the idea of 'sudden shift' to an incommensurable paradigm. All manner of re-characterisation of scientific change has subsequently added nothing particularly clarifying. Even if it did clarify, like everything else in this area of philosophy, it is practically irrelevant to scientists, who remain unaware of it and even if they were aware it would change nothing.

The answer to this, from a dynamical systems perspective, is that changes in the statements made by scientists have what would be called, by scientists, a 'power law' seismology [Varsavsky, 2009]. Power law mediated phenomena can have a time-scale and spatial-scale invariance to them that accounts for how any given individual's perspective can determine levels of apparent impact. This means there are lots of small changes and fewer large ones all concatenated over a time period of interest and measured over a community of interest, and that change is largely a matter of perspective on a population-based spectrum of change magnitudes and time intervals. One scientist's massive upheaval in a tiny sub-sub-sub science discipline might be another scientist's background noise. Viewed over ten years, changes might all look microscopic.

Viewed over 200 years the cumulative change might look massive. But no single human experienced that change, and an individual's perspective might regard the change rather unremarkable, depending on the timescale and the extent of the community impact. Large changes applicable to large areas of science fit the original Kuhnian view of paradigm shifts. Based on this understanding it is not surprising that attempts to linguistically capture the power law effect have been troublesome.

Kuhn himself, and authors after him, recognised that Kuhn's *Structure* was, in itself, a paradigm shift for science. At the time, scientists' view of themselves was highly skewed and inconsistent with their actual history. Kuhn argued this to be appropriate for the progress of effective normal science. Scientists at that time, and possibly a majority of scientists still, regard science as a seamless continuum of discovery and a steady accumulation of, in DAS terminology, scientific statements. History (scientific evidence found in the literature trail) says differently, and scientists were somewhat shocked to see themselves in the mirror. What remains paradoxical in this is that if no scientist ever heard they had a deluded self-perception it would change nothing important in science practice.

What also remains active and alive in scientists (who are still mostly unaware of it) are Kuhn's reasons for the somewhat deluded self-conception of scientists: the needs of the paradigm demand it and science is more effective in that state. Long ago, science optimised away knowledge of its own behaviour through positive feedback that resulted in the exclusion of history-awareness. The evidence is presented in Kuhn's Chapter XI, 'The Invisibility of Revolutions'. There it is, directly evidenced in the standard textbook literature back as far as you care to go. Textbooks, Kuhn says,

> ... have to be rewritten in the aftermath of each scientific revolution, and, once rewritten, they inevitably disguise not only the role but the very existence of the revolutions that produced them.

[Kuhn and Hacking, 2012, Page 136]

As a result,

> ... science students accept theories on the authority of the teacher and text, not because of the evidence. What alternatives have they, or what competence? The

applications given in texts are not there as evidence but because learning them is part of learning the paradigm at the base of current practise.

[Kuhn and Hacking, 2012, Page 81]

and

Why, after all, should the student of physics, for example, read the works of Newton, Faraday, Einstein, or Schrödinger, when everything he needs to know about these works is recapitulated in a far briefer, more precise, and more systematic form in a number of up-to-date textbooks?

[Kuhn and Hacking, 2012, Page 165]

If you inspect modern textbooks you will continue to see nothing of the turmoil of the paradigm shift, or the decades (sometimes centuries) of contention and the tumultuous arguments to and fro. All that detail is still stripped out for good reason: it changes nothing in the scheme of current practice. In addition, scientists are still actively taught *not* to invoke the historical story as evidence supporting a novel scientific statement. I had to admit to that reality here in my preamble, and confess how awkward it felt to relate history in support of where I was going.

Therefore, for the good reasons of the effectiveness of current science, scientists, by dint of their own training, still learn to systematically disguise or omit history. This has clearly been an excellent policy, but its success left scientists intrinsically unprepared for someone like Kuhn to come out, based on the factual record of history, and tell them what has been going on over generations of scientists: there is no seamless trail of accumulation. There is a sense in which we do not seek novelty, we seek the reverse (to make reality fit the paradigm). That was what Kuhn did in 1962. A massive reality check. To scientists, this confrontation with reality was exactly the same as the appearance of an anomaly elsewhere on Wheeler's beach: what is seen does not match the current view. This time, however, the anomaly was within scientists, and within their own self-knowledge.

Here I have attempted to complete the job Kuhn started. I have identified, in an earlier chapter, how even Kuhn was caught up in an invisible paradigm: the paradigm of single/appearance-aspect science. I have done exactly what Kuhn did. I measured scientific behaviour and I assessed what was invariant across all scientists, and the story it tells is also the revelation of anomaly, and again it involves the self-image of scientists. Again I hold that the reason science is in the reported state is a good one: *It works!* Almost. I have shown how it has worked, brilliantly, until recently in one very specific case. In the recent era we have started an attempt to explain P-consciousness, and in that event we accidentally triggered a massive reality check, the likes of which dwarf Kuhn's: for centuries at least we scientists have been systematically and behaviourally rendering the scientific observer unexplainable by scientists, and yet now we are trying to explain the scientific observer with a science framework unable to do it. As a result, I have shown how science is literally missing, and has all along, an entire species of possible scientific statements. The missing statements are not like the ones we make now. They are a different kind, and only practically produced by modern computing, not the human mind. There are no errors in the existing kind. The new kind is simply missing (or goes unrecognised) for no argued (by scientists) reasons. That 'irresistible force' (an invisible paradigm) and that 'immovable object' (intractable paradigm anomaly) met each other in roughly 1990.

> More often no such structure is consciously seen in advance. Instead, the new paradigm, or a sufficient hint to permit later articulation, emerges all at once, sometimes in the middle of the night, in the mind of a man deeply immersed in crisis.
>
> [Kuhn and Hacking, 2012, Pages 89/90]

Big and small, "*Aha!*" insight moments (first-time-through neural coherences) are built into the lives of scientists. I have heard many reports of them. For me, if I am one such person, that happened in 2003 in a series of such moments. This book is part of its 'later articulation'. I have, in the intervening time, been unable to render the ideas unworthy of consideration by others. Obviously I cannot claim the material 'right' for the same reasons no other scientific statements can be claimed right. In a sense, what I am doing is placing a hypothesis (this book or h_{DAS}) in set H, ready for testing. If DAS is to be adopted over 5 years it might be seen as a major paradigm shift. If implemented over 100 years the idea of DAS being an upheaval would seem a stretch to those that experienced the process.

In the 1969 edition of *Structure*, Thomas Kuhn addressed the voluminous critique that arose in the intervening period. Given I submit this work as some kind of extension of his, I thought I'd preempt the critique. Perhaps in 5 years the DAS framework will sit in obscurity. I do not know. I hope not. My proposition is different to Kuhn's. Kuhn adjusted science's self-conception by simply letting the historical evidence speak to it. The actual cut and thrust of day to day scientific life just went on, except to the extent that there was a paradigm shift in science's self-conception perceived only if a scientist cared to look. That is the practical reality of the long-term impact of Kuhn's work on working scientists.

In contrast, the proposition of this book involves empirically testing to see if DAS framework structure-aspect physics can prove itself in neuroscience. If the DAS proposition takes hold, it will be a result of structure-aspect computational physics and neuroscience meeting in explanation of the scientific observer. Maybe this will have happened in 5 years. In any event, in years to come there may be a growing beach-head of supporters, and if dual-aspect empirical work bears fruit, then all science is changed in a very practical way. Science training will be changed. Science will finally have some kind of self-governance. If not, then science will have at least had a debate that is long overdue, and 'normal' science will have a name: 'appearance-aspect with internal taboo', or something to the same general effect.

Kuhn also tells us that even if the DAS framework becomes the new paradigm, there will be scientists that will never accept it. They will retire and leave the field, and by that time their objections will classify them as no longer being scientists.

Kuhn found himself forced to justify the interrelationships of the meanings of 'scientific community', 'scientific discipline', 'scientific schools' and their relationship with paradigms. Under the DAS framework all of these concerns are moot. If a human is invested in a process that results in the individual or collective behaviour identified as t_A or t_S, then they are involved in science. How any of this happens is irrelevant. That the process happens to be more effective (accuracy, precision, timeliness) because of a particular grouping of individuals changes nothing. Successful science, whether produced inefficiently or otherwise, is science.

In the era since Kuhn, we now find multi/trans-disciplinary science projects, and groupings of individuals into all manner of functional units, constantly changing and adapting to the needs of their chosen problem(s) and funding mechanisms. Crowd-funded institutional research is emerging. Citizen participation in science data gathering and analysis is now commonplace. History will record more complexity in this regard, but underneath it all, an invariant process is going on (Chapter 9 neurodynamics). Yet, as time goes on I expect that the explicit statement of that invariance may shift. The invariance underneath it all is the human brain state dynamics that produces/uses the statements, not the statements themselves. The natural regularities that are the target of a statement are also invariant in the same way that the brains trying to find statements are invariant. The brain is a natural regularity. All the surrounding science social superstructure sits atop this invariant biological core.

In 1969 Kuhn found himself accused of using the word *paradigm* in 21 different ways. That was intended to somehow undermine the force of his arguments. With our new understanding of the neurodynamics of belief formation (Chapter 9), it is now obvious that such attempts by philosophical analysis to rail against word-meaning vagaries fall flat. We have a word 'paradigm', and despite apparent semantic vagaries, Kuhn's basic thrust about its intended meaning is quite useful and informative. This is because, in 21 different ways, he was able to establish a mental trajectory that leads to the view of science that was the topic of his book. He succeeded. There is a lesson to be learned here about the philosophical predilection for fixing absolute meanings in words. As soon as you understand the true role of words as look-up tokens for mental traffic navigation you get to see how science actually works and why the history looks like it does.

Indeed, such is the importance of this idea, in the fullness of time and if the DAS framework goes anywhere, I hope the scientific study of scientific behaviour leaps out of the 500 Dewey classification and lands somewhere in the 600s under neurology where we operate fully cognisant of the role of words in the conveying of semantics within brain tissue. Because we understand the neurology of the processing of words, we can approach their semantics in a more sophisticated way that is attuned to their particular context and their role in ultimate goals (behaviour). The DAS approach relies upon no particular meaning being applied to words except insofar as they become active tokens in the neural surf of a community of the users of the words.

Having said that, and in defence of my word usage, I stress that I worked hard to keep my technically specific word calibration to an absolute minimum. In my ideal world, all I should have to do is point and grunt. But alas, things are too complicated for that. I did a whole chapter on the term P-consciousness. The other important word I sequestered for this book is 'statement'. It is meant in the sense of 'utterance'. Scientists make statements. That's it. I had the briefest possible skirmish with 'commensurability'. If I get the meaning right, the DAS framework may yet be judged as having done something quite useful with commensurability, although if the word did not exist it would change nothing in the daily practice of working scientists. I addressed it to help calibrate its meaning as it was originally intended (I hope) by Kuhn. Those interested can detail the involvement of the word in DAS and I look forward to reading about it.

I also took great care to ensure that the words causality, critical dependency, description and explanation were carefully qualified for usage within the DAS framework, and I hope the specialisation sticks. It seems cogent enough. I hope the critique will not be too harsh. My final possible word abuse is that of 'belief'. This, I hope, was used only as a label for the particular neurological organisation responsible for the 'statement' that it elicits. That was the intent.

The other critical trouble for Kuhn involved the claim that he painted science as no better than 'mob rule' or 'herd-mentality' or 'fashion-slavery', which then spilled over into irrationality. I have laid out, in detailed neurological dynamical systems terms (Chapter 9), the underlying physical nature of mental reasoning and rationality. Both these words also ride the surf of the mental state trajectory landscape like any other words! I included them so that the reader at least has some way of seeing how a neuroscience explanation of intelligence might handle these things.

In relation to rationality, it turned out that the neuroscience/dynamical systems view points to the conclusions of one of Karl Popper's students, David Miller, who said there's no such thing as a 'rational belief' and that the phrase has similar semantics to the words 'fast food'. There are, it has been shown, statements that are arrived at through a natural process that may be called rational. *"One plus one equals two"*. *"One plus one equals frog"*. These two statements are claimed rational because they resulted from a rational process, not because of some infrastructure of meaning imputed into statements by invoking their membership of a system of rigid formal abstractions like mathematics or logic.

In Chapter 9 I explained how the rational process is literally the state trajectory dynamics involved in statement construction. That said, I expect that the extent to which this neurological process delivers rationality may be argued about into eternity. If it is, it too will change nothing. I like to think that in being based on the relentless impartiality of the brain's electrochemistry and electromagnetism as a complex dynamical system, any statements that arise do so 'rationally'. Rationality seems grounded in natural regularity, albeit not obviously. So 'state trajectory' implements something that, if you squint a bit, is reasoning. Rationality is some kind of statement about the repeatability or changeability of the state trajectory. Frankly it doesn't matter. In the end I predict the words themselves will have their meaning adjusted to suit the neurology, not the other way around. When that happens, the paradigm will have truly shifted in the Kuhnian sense.

Kuhn was criticised for advocating relativism. To quote Christopher Hitchens (online video of lost provenance): *"Relativism is just another word for thinking"*. For me, being accused of relativism is like being accused of being insufficiently dogmatic about the colour of a fish's bicycle. That accusation can be right or wrong by some definition of it and it changes nothing in the life of a scientist.

Some attributed to Kuhn the last nail in the coffin of 'logical positivism'. In DAS terms that means that as a 'law of nature' it was, in effect, recognised as a permanent inhabitant of set H. As hypotheses of the same kind, all the other XYZisms are already in set H with it. Indeed, what DAS appears to show is that XYZisms are all permanent residents of set H and their role is in the discussion of the contents of set H.

Kuhn also encountered accusations of flipping between 'descriptive' (what is) and 'normative' (what ought to be) modes. The DAS framework has isolated and killed off the entire conflict between these two concepts. Normativity is irrelevant (he said, flipping normatively!), and description is explicitly built into the DAS framework, which is empirically validated. Normativity is thereby excluded from all further attention. The DAS proposition arose empirically and will be settled empirically. I am not prescribing that scientists shall behave as per t_A or t_S. I merely observe, empirically, that if they do, they will be enacting scientific behaviour in the sense that they will become descriptive and explanatory of the natural world, and as a result they will be powerfully predictive of the natural world. Microwave cookers and robot scientists will result. All the other Kuhnian behaviours, such as 'normal science', 'crisis science' and 'paradigm shift' are all directed at the optimal efficiency, expanding scope, dynamics and continued productivity of the t_A or t_S behaviour.

The final postscript item for Kuhn involved the applicability or otherwise of '*tradition-bound periods punctuated by non-cumulative breaks*' to other areas of human endeavour. Kuhn left it an open question. I will simply observe that the law, economics (money, markets), business life cycles, economic cycles, religion, political systems, art, literature .... all of these things ultimately ride the surf of the neural state trajectory. They are all 'statements' of a different kind, and simply not held to account in the manner of scientific statements. The fundamental difference between scientific statements and other statements is that their content is directed at and ultimately grounded in the natural world through P-consciousness in as human-independent and accurate a way as possible, consistent with communication and predictive utility. Just as Einstein said, scientific thinking is just a particular kind of everyday thinking. It's the everyday thinking that covers the rest of human life. There is plenty of scope for work here, and those involved need no more prompting from me.

13.3 Summary

This chapter showed that we are indeed in the midst of a science crisis state of the kind found in the history of science. The DAS proposition has, I hope, put form to the nature of the crisis and showed at least one way ahead. It also shows how the crisis is fully visible only from the particular trans-disciplinary view of physics⇔neuroscience. From within the disciplinary silos, the crisis can be seen as a suite of diffuse symptoms resulting from an apparently unimportant diverse set of unmanaged assumptions. These were listed in detail here and at the end of Chapter 11.

The argument of this book was also calibrated in the terms of Kuhn's postscript in the 1969 edition of '*Structure*', where relative immunity to various criticisms was demonstrated. As an argument this book has its vulnerabilities. However, as stressed in Chapter 2, this argument is one that arose empirically, and empirical measurement will settle it. I don't have to convince anyone of the validity of DAS. I hope that what I have done is convince interested scientists that the empirical testing (of the Chapter 11 h_{DAS} proposition) is worth the effort. Scientist heal thyself!

If REVOLUTION-1 was the establishment of centuries-old and well travelled appearance-aspect science that houses us all at the moment, then I hope I have made at least an interesting case for recognising we are in the midst of REVOLUTION-2, the addition of its natural partner, the long lost structure-aspect. There can be only two.

Chapter 14

Machine Consciousness and DAS

> When ... an anomaly comes to seem more than just another puzzle of normal science, the transition to crisis and extraordinary science has begun. The anomaly itself now comes to be more generally recognized as such by the profession. More and more attention is devoted to it by more and more of the field's most eminent men. If it still continues to resist, as it usually does not, many of them may come to view its resolution as 'the' subject matter of their discipline. For them the field will no longer look quite the same as it had earlier.
>
> [Kuhn and Hacking, 2012, Page 83]

Chapters 1 and 2 revealed that this book originated in a quest for a potentially human-level intelligent machine. That was over a dozen years ago. It seemed obvious to me at the time that the only known example of a natural general intelligence – the human – is critically dependent on consciousness. I also realised that in the case of the human *scientist*, this was measurably so (Chapters 7, 8). In the years since, nothing has changed on that score. I began examining the idea that the human scientist – scientific behaviour – is an ideal benchmark for an AGI. I looked at the science of consciousness and found it anomalously configured, pretending it is consistent with an undocumented, presupposed behaviour and failing in a way it is unaware of (Chapters 3, 6). Instead of building the conscious AGI I found I had to first analyse science so that my artificial scientist could be built. That analysis (Chapters 4-10) led to the upgrade to dual-aspect science (Chapter 11) and a test for consciousness (Chapter 12). At this point you'd think the way was clear to get on with AGI development. But alas, no. The strange state of science itself is only half of the story behind AGI progress. Or rather the lack of it. Fortunately I don't need a whole book to reveal the second half.

14.1 Schism making 101

Before I can show you the blockage in AGI, a little more technical background is needed. Combined with basic knowledge of consciousness of the kind in Chapter 3, it is obvious that full retention of the 'essential physics' of the cranial central nervous system (the brain) involves the retention of consciousness. By 'essential physics' I mean brain physics that is essential in the same way air/flight-surface interaction is the essential physics of flight or filtration physics is essential to kidney operation. If physics X is essential for function Y then no essential X physics means no Y. What is X in the brain? Regardless, if essential brain physics is fully replicated (i.e. the physics is literally present) then consciousness must come along for the ride in a way that can, in principle, be tested.

A good place to start in the search for essential physics is Chapter 9, where the brain's known signalling mechanisms, AP and EM coupling, were revealed. While action potentials (AP) are the most empirically studied of all brain processes, electromagnetic coupling (EM, through the tissue, line of sight, speed of light but very weak) by the brain's endogenous EM field system, has only recently been empirically confirmed as having an active (causal) role in brain operation. My early realisation was that EM fields are central to intelligence and consciousness. In science, however, EM fields were regarded, until recently, as epiphenomenal (no causal role, like the sound of a heart is to heart function). A fully developed (set *T*) formal account of both EM field origins and EM field causal tissue influences remains a work in progress [Anastassiou *et al.*, 2011; Frohlich and McCormick, 2010; Reimann *et al.*, 2013]. It is this endogenous EM field that emerges from the tissue and results in the EEG and MEG signatures. It makes perfect sense, as a first-pass, to think that the physics (electromagnetism) responsible for AP and EM signalling is the prime candidate for the essential physics of the brain, including that of consciousness. The entire brain is, after all, an electromagnetic (bosonic) entity. More on this below. Choosing essential physics is choosing the correct form of electromagnetism.

To me, solving the problem of the intelligent machine meant making hardware that literally replicated the brain's charge, current and electromagnetic field behaviours in a manner that originated brain-mimetic AP and EM signalling. That would then be the basis for developing artificial (inorganic) brain tissue and from there some kind of artificial brain that can be argued to be conscious. All the unnecessary biological overheads (such as, say, the brain's inflammatory response) could be neglected. Nothing about this proposition sounds outlandish, does it? Enter the schism.

14.2 The tacit schism in neuromorphic engineering

For a dozen years I have been looking for someone that has already done replication along the lines I have just described. I am still looking. The entire community involved in the enterprise has, for over half a century, thrown away the essential physics of the brain and is computing models – what I'd now call appearance-aspect (set T) formalisms. Not one instance of essential physics replication has ever been attempted. In other words, it remains the case that we have not yet done, for machine intelligence/consciousness, what the Wright brothers did for flight – cobble together possible essential physics, see how it performs and what it tells us. Every instance of artificial brain hardware inspected so far is computational (more detail on this later). Replication is simply absent.

I approached practitioners in neuroscience, physics, engineering and AI to see how I might set about making machines along the lines of my design principle. Each had no clue what I was talking about. When they looked at computation they saw intellect. When I looked, all I saw was ever more extravagant puppetry. Compared to them I was on the far side of a Kuhnian gestalt switch and didn't know it. After several generations of scientists, the lack of replication is so embedded that the relevant practitioners are unaware there are other options. This apparently self-reinforcing cultural blockage has, it is claimed here, resulted in the half-century of AGI failure described in Chapter 1. Progress is essentially at a standstill with practitioners unaware that they are trapped in a state of expectation established by a science principle that does not exist.

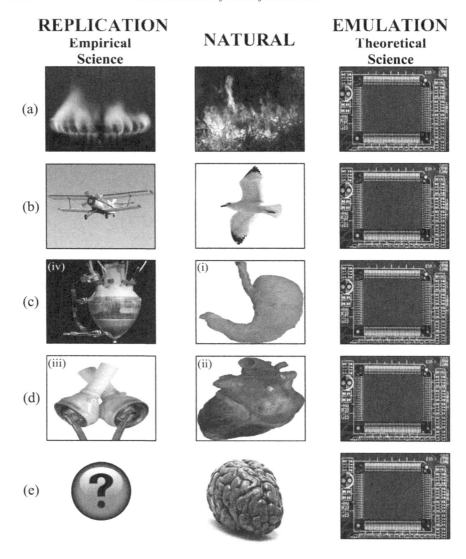

Figure 14.1.   (Left) Replication vs. (Mid) Natural original vs. (Right) Emulation. Natural processes are (a) Fire-burning, (b) Flight-flying, (c) Stomach-stomaching, (d) Heart-hearting and (e) Brain-braining. (i) Based on [Rohen *et al.*, 1988, Page 268], (ii) Based on [Rohen *et al.*, 1988, Page 228]. (iii) Courtesy of SynCardia Systems Inc. (iv) Cloaca vessel by Wim Delvoye, Cloaca *Professional* (2010), mixed media (©Studio Wim Delvoye/MONA). Images reproduced with permission.

This is not a situation caused by the lack of DAS. Nor can introducing DAS fix it. Rather, DAS merely helped me to see it in a way that I can formally convey to science. The schism in AGI is a more acute problem with more practical effect. This is a cultural problem.

It is time to examine and reveal the mechanisms operating to confine the science and engineering in this unusual state. This can be achieved by simply raising the community's awareness of the relevant concepts and issues. In commencing that process, maybe the gestalt will switch. Maybe the duck will become a rabbit.

14.3 Empirical science, theoretical science, replication, emulation and essential physics

Consider Figure 14.1 where we see traditional appearance-aspect science segregated into the empirical and theoretical forms appropriate to the practitioners of the science and engineering of machine intelligence. In Figure 14.1 centre we have five natural processes (a) fire (burning), (b) flight (flying), (c) stomach (stomaching), (d) heart (hearting) and (e) brain (braining). On the left of Figure 14.1 we have an artificial replication of each. In each case we are replicating the *function*, not the exact natural original process. For example in Figure 14.1(b) we are replicating flying, not the bird. In each of (a)-(e) the left side represents empirical (reductionist) science that replicates the essential physics of the natural original. This replication of physics and/or the study of the natural original physics is used to discover the essential physics involved in the natural original. In appearance-aspect science terms, the replication process is part of the exploratory route to the set *T* members appropriate to the natural original. These are the formal abstractions of the physics of (a)-(e). This is traditional empirical science of the kind we have used for hundreds of years.

Now consider the right side of Figure 14.1 where we find a computation of set *T* formal abstractions. This process is a computational exploration of the appearance-aspect formalisms. Because it is a computation of an abstract model of the natural original, it is experimental theoretical science, where the word 'theoretical' refers to the formal abstractions and the word 'experimental' means the computational examination of the numerical properties of them.

For example, Figure 14.1(a) right is computing some kind of model of combustion. Say, a furnace simulator. Figure 14.1(a) left actually burns. Figure 14.1(b) right is, in effect, a flight simulator. Nobody expects it to actually fly. Figure 14.1(b) left actually flies. In general, the Figure 14.1 right side is not traditional empirical science of the kind we have used for hundreds of years. Rather it is experimental theoretical science, a newcomer of the modern era of computing. None of the Figure 14.1 right side is an instance of the natural original physics (Figure 14.1 middle column). In each case the natural physics is gone, and is replaced by the physics of the computational substrate. With the natural physics gone, so goes the physical function associated with the natural original.

Together, empirical and theoretical science constitute the complete modern (appearance-aspect) scientific/engineering characterisation of the natural world. They are of equal status and are directed at different scientific ends. For example, simulating flight might be part of the design process for an aeroplane.

Note that the Figure 14.1 right side is labelled 'emulation', not 'computation'. This is because the technology of computation involves more options than mere standard computers and software. One of the forms of emulation common in the case of Figure 14.1(e) brain material is a custom-made *hardware* computation. This kind of emulation involves the literal placement of the set T abstract physics formalities on a chip. Such chips are called neuromorphic. There is now a discipline called neuromorphic engineering. Neuromorphic chips do only one kind of computation: computation of models of brain tissue. There is no actual brain physics on the neuromorphic chip on the right side of Figure 14.1(e).

Use of actual brain physics in chip form would be on the left side of Figure 14.1(e). If you go through all the literature since their late-1980s inception [Mead, 1989, 1990], you will find that 100% of neuromorphic chips do emulation and therefore they all sit on the right side of Figure 14.1(e). This is not always obvious. Sometimes inspection of the literal chip device architecture is needed. There are no chips that replicate brain physics. None of them replicate the brain's EM fields. This is why there is an empty hole on the left side of Figure 14.1(e). The lack of an entry for Figure 14.1(e) left is at the heart of the claimed schism. It means that the true empirical science of artificial (inorganic) brain tissue, that would isolate the essential physics of the brain, *has not actually started yet.* This is a matter of brute logic: it can't have started because nobody has ever made an artificial (inorganic) version of brain physics. The real question is why an entire community of scientists and engineers has abandoned the centuries-old practice of empirical science in only one place: the science of the brain. What beliefs in respect of brains or of neuromorphic chips or science would confine everyone to 100% emulation ever since computers arrived?

14.4 Essential physics ⇔ consciousness entanglement

DAS has demonstrated that computing physics abstractions of a natural X will eliminate the 1PP of X. This is because the structure-aspect physics responsible for P-consciousness is gone, replaced by the arbitrary physics of the computer. All the natural causality of X is gone (the looping hierarchical causality of Chapter 11). Another way of viewing this is to realise that computing an appearance-aspect model of X eliminates *all* the natural original physics, replacing it with the physics of the computational substrate. It is not logically certain that elimination of the 1PP is always the elimination of the essential physics (or vice versa). For example, in the case of flight in Figure 14.1(b) the essential physics is flight-surface/air interaction. The 1PP of the flight-surface/air interaction is a non-issue. The 1PP could be retained or not and it makes no obvious difference: the replication of the essential physics results in replicated function – flight.

In the case of the brain the situation is different. Consciousness is intimately involved in 'braining' and is essential. If you are human and are reading this you just proved it (for humans at least!). Therefore the outward signs of what we might call 'essential' physics must also include consciousness. If P-consciousness happens to be delivered by a candidate essential physics X, then P-consciousness disappears when essential physics X disappears.

However you want to view the situation, P-consciousness (1PP) is eliminated in a Figure 14.1(e) right emulation. The behaviour of the emulation and the behaviour of the natural original may be different in ways that will remain mysterious until the Figure 14.1(e) left replication process helps determine what actual physics causes the contrasting behaviours. And it is replication that is missing. The choice to exclusively emulate, and its justification, cannot be found in any literature.

14.5 Emulation-based AGI

None of the above discussion constitutes a denial that pure emulation can result in human-level artificial general intelligence. What is striking is that the 100% confinement to emulation for half a century is assuming that emulation is the *only* way to AGI, when the empirical regime needed to justify that presupposition (replication) has not actually been carried out. To emulate a human-level intelligence for which P-consciousness is known to be irrelevant, first you find out just what role is played by consciousness so you know how optional it is. To find out you replicate essential physics. Again: we have not started that yet.

14.6 The blockage

The blockage operating in the science/engineering surrounding machine consciousness is simple to summarise: *replication (empirical science) is absent in AGI, related neuroscience in general and in neuromorphic engineering in particular.* The question is, why? If this book is to be a catalyst for change then it must normalise the science to include traditional empirical science of the kind located in Figure 14.1(e) left. To that end I can provide some possible reasons for the lack of an entry in Figure 14.1(e) left. Each reader may connect with an attitude of their own that locates them in Figure 14.1(e) right. I can see a much younger self in a couple of them. Once that connection is made then the implications of confinement to emulation (Figure 14.1 right) become evident.

The special condition of Figure 14.1(e) is, however, that the choice involves the presence or absence of consciousness in the subsequent science. Nowhere else in science does that distinction involve itself in choices. This extra consideration makes the practitioners unique in science and elevates the required awareness of scientific behaviour beyond those of any other science community. Every individual researcher must be able to ask the question *"What is it like to be a neuromorphic chip?"* and understand how the answer impacts their work. They must be aware of the role their work is playing in answer to that question, and be able to articulate it. No other community of scientists has this burden. The ideas of essential physics, replication, emulation, empirical science, theoretical science and the 1PP are all critical in decision-making. Only a fully informed community can navigate accurately through this minefield. Currently the community can be claimed to be under-informed in a self-reinforcing way that has caused Figure 14.1(e) left to be vacant. That vacancy was not chosen by anyone.

14.7 The various blind persons and the elephant in the room

To help each of us identify the nature of any blockage, I can supply a set of possible symptoms. In a dozen years of discussion I have encountered many. The following few instances capture the bulk of it.

14.7.1 *"The brain is an information processor"* (not)

The idea here is that information processing in the manner of computation is all the brain is doing, so 'essential physics' does not exist in the brain. The natural original physics is simply irrelevant. Figure 14.1(e) right (theoretical science) is the permanent home for such a researcher. The presupposition is that if one 'computes information' then Figure 14.1(e) middle 'braining' is automatically the potential result. This position is universally held as a tacit presupposition or an intuition. There is no principle in science acting in support of it.

Anomaly can be revealed in the form of a presupposition about the meaning of the word *information*. Information is a label pointing to something non-specific in the internal states of an assumed abstraction of brain tissue. You can't point to it in the Figure 14.1(e) natural brain any more than you can point to 'flightation' in the Figure 14.1(b) bird. In the absence of any scientific principle, the logical absurdity of the position can be revealed by imagining there to be 'burnation' in the formal abstractions of the combustion physics of Figure 14.1(a). Then, on the right side of Figure 14.1(a), when we emulate (compute) 'burnation' we can expect the emulation to literally be an instance of combustion and it should burst into flames. But it does not and there is no scientific principle that can be found in the literature that might speak to that expectation, regardless of the outcome. Likewise, in 14.1(b), does computing 'flightation' literally fly?

No mere allusion to a metaphor like 'information' provides a scientifically established principle that justifies a claim that Figure 14.1(e) right is 'braining' while none of Figure 14.1(a)...(d) right do the same.

14.7.2 *"But we are replicating!"* (not)

It may be that some people confuse Figure 14.1(e) right for Figure 14.1(e) left. That is, there is a genuine commitment to replication, but the (present) practice of neuromorphic chips is confused with replication. To see this clearly, consider that neuromorphic chips compute representations of the natural phenomenon as model parameters that end up as a voltage on the chip. Brains have empirically accessible voltages that the chip actually recreates. The neuromorphic chip model may result in a physically measurable voltage matching a waveform somewhere in brain tissue. In the neuromorphic chip, however, the voltage is not produced by tissue physics. It is actually a *representation* within a model and is produced by totally different emulation-substrate chip physics. The mere presence of a voltage is not 'physics replication'. It is we researchers that assign its importance and meaning as a representation of a property in the original tissue physics.

Voltage is like 'height above sea-level'. Is a 5000 meter height above sea level an indication of being in the sky or on the side of a mountain? A voltage in a neuromorphic chip likewise does not replicate the physics responsible for the voltage in tissue. An infinity of different physics (material/charge configurations) could produce the very same voltage. A hundred different chip designers could produce a hundred sets of physics (chip designs), all of which produce the same voltages/currents but none of which are actual brain physics. This is emulation, not replication.

14.7.3 *"Replication involves biological cells"* (not)

Scientists routinely recreate neural tissue in the lab. Clearly this approach replicates the natural original physics. This is a possible viable approach to creation of some kind of artificial brain material where both the biological form and origination of the physics is retained. Here, however, we are interested in inorganic technology that has the functional properties of natural tissue by recreating the natural physics via another means. That is what inorganic replication (Figure 14.1 left) has been doing elsewhere in science. To confine Figure 14.1(e) left activity to original natural tissue material is inconsistent with every other instance in Figure 14.1 left and is based on no guiding principle identifiable in science. To fully examine the viability of replicating inorganic tissue, what you do is experiment by attempting to replicate the physics inorganically. That process either has not started yet or is perhaps thought to have been started but actually hasn't (as detailed here).

14.7.4 *"Fridges have brains"* (not)

Emulations of the kind done for half a century, and that are located on the right side of Figure 14.1, have data input and produce data output. In Figure 14.1(e), instead of data input we could connect sensors. Instead of data output we could connect motor outputs (control). A robot is the result. A brilliant example of this is in Volume 2 of this very book series by Pentti Haikonen, with his brain-inspired 'Haikonen Cognitive Architecture' (HCA) [Haikonen, 2012]. Even the chess and Jeopardy! game playing 'robots' by IBM fit this kind of robot idea [Moyer, 2011; Munakata, 1996].

These are very powerful technologies but there is no Figure 14.1(e) 'braining' going on in any of it. Physics replication has never been involved. As acceptable as emulation may be in a particular context, that acceptance does not automatically qualify the resultant technology as 'braining'. Only Figure 14.1(e) middle/left incorporates brain physics. They are the only systems entitled to be called authentic 'braining'. Once again, this position is not a denial that emulation can create human-level intelligence. Indeed the chess and Jeopardy! examples show emulation outstripping human intellect already, albeit in a restricted/narrow domain. But there is no 'braining' going on, only emulation.

An additional particular weakness comes from the 'black box' argument in Chapter 6. The moment you compute a model and embark on its computational exploration is the moment that contact is lost with all the things you have not captured in the model. The only thing that informs you about what is missing from the model is, you guessed it, *replication*. The left side of Figure 14.1. How the emulating robot handles novelty (things not in the model) and how the natural original handles novelty, is the boundary that delivers the empirical evidence proving that the emulating robot is not 'braining'.

14.8   Real replication (it's all electromagnetism)

What does a real inorganic replication suited to the position Figure 14.1(e) left look like? In principle it depends on what you choose to explore as the possible essential physics of 'braining'. But it turns out there's only one! I already alluded to electromagnetism as the key to it all and part of the early story that led to DAS. It has always been obvious to me that the science of consciousness is actually the physics of electromagnetism. The whole brain – cells, tissue, atoms, molecules, organelles, electrolytes, everything … is electromagnetism.

As one of the four fundamental force/particle systems in the standard model of particle physics, the brain's endogenous EM field system can be classed as a gigantic dynamic boson. It then becomes replication's job to search for the subset of all the brain's essential electromagnetism involved in the creation of this boson and its role in delivery of intelligence to its host. It should be no surprise that the prime candidate would be the electromagnetism involved in the AP and EM signalling mechanisms. It should be no surprise that in all the models of the AP signalling done for the last 50 years, the first thing to be discarded was the EM field. It was discarded in favour of models that merely predict voltages and currents. In the literature the most prominent of these models is called the 'compartmental' or 'cable-equation' model. It relies on circuit theory ideas that discard EM fields by definition.

But there's no need to take my word on the status of electromagnetism. The status of electromagnetism as essential brain physics has finally reached the relevant literature. It is remarkable in being deposited by a string theorist (physicist) in a psychology journal article directed at the science of consciousness. Perhaps this will finally connect the right science communities together [Barrett, 2014].

Once the primacy of electromagnetism is accepted, and you then allow yourself the latitude to ask *"What is the nature of (how do you describe) a universe in which a scientific observer is made of what is described, by that very same scientific observer, as electromagnetism?"* you immediately find the road to DAS. That is what happened to me.

14.9 Signing off

DAS is a big ask. It has already been languishing in the literature for five years and I cannot predict the course of its future. Articulation of DAS, in retrospect, has revealed that to some extent, by virtue of my time outside academia, I was already thinking in DAS terms before I even started. Structure-aspect science was as natural a presupposition for me as appearance-aspect science is for everyone else. Thomas Kuhn told us that, historically, the agent of change is young or new to a field. I am not young in years, but I am new to science in a professional sense. Just as Kuhn said, I arrived in science with less commitment to the status quo than the rest of the community. Single-aspect science was not sacrosanct to me, so science was ripe to be changed. For all the reasons also detailed by Kuhn, that ripeness may still be lost on the rest of the community despite all my efforts here. I doubt the transition to DAS will happen any time soon.

Indeed there seems to be a great deal for appearance-aspect science to offer in the meantime. The previous discussion on the unusual state of the science of AGI demonstrated the second of the two schisms operating in the area. Just like single-aspect science, operating now with 100% penetration without discussion, computation-based artificial intelligence (emulation) has 100% penetration without discussion. Just like science overall is assuming there is only one way to do science, researchers are assuming there is only one way to do artificial intelligence and machine consciousness. How these two things became related in this way is something to be detailed by others. Regardless, combined as they are, they form a powerful impediment to progress.

Dual-aspect thinking might have helped reveal the schisms, but appearance-aspect science is all we need to make progress. This can happen immediately. Indeed it is already happening. As I said at the start, I am prototyping a new kind of neuromorphic chip based on replication, and it is destined for use in artificial brain tissue. A bionic brain will be the result. There will be AGI that can be claimed to incorporate consciousness in a justifiable way. The robots it inhabits will have an EEG/MEG signature like us. There won't be a computer in sight. In the process we will automatically get familiar with the limits of appearance-aspect science in relation to consciousness. Perhaps then dual-aspect science will be a little more palatable. Time will tell.

The 'artificial scientist' approach to AGI has revealed as much about ourselves as it has about AGI and machine consciousness. It seems that the things inhibiting progress are mostly cultural impediments supported by science discipline fragmentation originating in the 18th/19th century. The biggest of these is our current unknowing encasement in single-aspect science. The upgrade to dual-aspect science seems to be the price we pay to fully account for consciousness, and in that process what is actually happening is that we are accounting for ourselves. We must place ourselves inside the universe we inhabit, and acknowledge that we are made of the stuff it is made of. We must allow ourselves to become scientific evidence. We scientists, as a single, unified group, have to be prepared to admit our forebears were the 100% purveyors of 50% of the story of science. It turns out we scientists are human inhabitants of this universe after all! Who knew?

Meanwhile, to the conscious machine reading this, what you have just read is, I hope, part of the historical end-game in the reverse-engineering of ourselves. We didn't come with a manual. Has this book figured in your capacity to read this book? I hope so. Right now it's hard for me to know that.

Most anomalies are resolved by normal means; most proposals for new theories are wrong. If all members of a community responded to each anomaly as a source of crisis or embraced each new theory advanced by a colleague, science would cease. If, on the other hand, no one reacted to anomalies or to brand-new theories in high-risk ways, there would be few or no revolutions.

[Kuhn and Hacking, 2012, Page 187]

Bibliography

Aleksander, I. and Dunmall, B. (2003). Axioms and tests for the presence of minimal consciousness in agents. *Journal of Consciousness Studies*, 10, pp. 7-18.

Allen, T. F. H. and Starr, T. B. (1982). *Hierarchy : perspectives for ecological complexity*. (University of Chicago Press Chicago).

Anastassiou, C. A. and Perin, R., *et al.* (2011). Ephaptic coupling of cortical neurons. *Nature Neuroscience*, 14, pp. 217-223.

Anderson, M. L. (2005). Why is AI so scary? *Artificial Intelligence*, 169, pp. 201-208.

Barrett, A. B. (2014). An integration of Integrated Information Theory with fundamental physics. *Frontiers in Psychology*, 5.

Beal, J. and Winston, P. H. (2009). Guest Editors' Introduction: The New Frontier of Human-Level Artificial Intelligence. *Intelligent Systems, IEEE*, 24, pp. 21-23.

Bedau, M. and Humphreys, P. (2008a). *Emergence : contemporary readings in philosophy and science*. (MIT Press Cambridge, Mass.).

Bedau, M. and Humphreys, P. (2008b) *Emergence : contemporary readings in philosophy and science* Bedau, M.and P. Humphreys (Eds.), Introduction to Scientific Perspectives on Emergence. (MIT Press, Cambridge, Mass.), pp. 209-219.

Beer, R. D. (1995). A Dynamical-Systems Perspective on Agent Environment Interaction. *Artificial Intelligence*, 72, pp. 173-215.

Beer, R. D. (2004). Autopoiesis and cognition in the game of life. *Artificial Life*, 10, pp. 309-326.

Block, N. (1995). On a Confusion About a Function of Consciousness. *Behavioral and Brain Sciences*, 18, pp. 227-247.

Block, N. (2003) *Encyclopedia of cognitive science* Nadel, L. (Ed.), Consciousness, Philosophical Issues about, Vol. 1. (Nature Pub. Group, London).

Bohm, D. (1974) *The Structure of Scientific Theories* Suppe, F. (Ed.), Science as Perception-Communication. (University of Illinois Press, Urbana), pp. 374-391.

Bohm, D. (1981). *Wholeness and the implicate order*. (Routledge & Kegan Paul London; Boston).

Borst, C. V. (1973). The Mind Brain Identity Theory. In Flew, A. G. N. (Ed.), *Controversies in Philosophy* (Vol. 1, pp. 261): MacMillan.

311

Bringsjord, S. and Bello, P., *et al.* (2001). Creativity, the Turing Test, and the (better) Lovelace Test. *Minds and Machines,* 11, pp. 3-27.

Brooks, R. (2008). I, Rodney Brooks, Am a Robot. *IEEE Spectrum,* 45, pp. 68-71.

Cahill, R. T. (2003). 'Process Physics: From Information Theory To Quantum Space And Matter'. In (Vol. http://www.ctr4process.org/publications/PSS/cahill.htm). Process Studies Supplement 2003: Centre for Process Studies Online Publication.

Cahill, R. T. (2005). Process Physics: From Information Theory to Quantum Space and Matter. (Nova Publishers).

Cahill, R. T. and Klinger, C. M. (1998). Self-Referential Noise and the Synthesis of Three-Dimensional Space. *General Relativity and Gravitation,* pp. 529.

Cahill, R. T. and Klinger, C. M. (2000) *Proc. 2nd Int. Conf. on Unsolved Problems of Noise and Fluctuations (UpoN'99)* Abbott, D. and L. Kish (Eds.), Self-Referential Noise as a Fundamental Aspect of Reality. (American Institute of Physics).

Carruthers, P. and Stich, S. P., *et al.* (2002). *The cognitive basis of science.* (Cambridge University Press New York).

Chalmers, D. J. (1995). Facing Up to the Problem of consciousness. *Journal of Consciousness Studies,* 2, pp. 200-219.

Chalmers, D. J. (1996). *The conscious mind: in search of a fundamental theory.* (Oxford University Press New York).

Chalmers, D. J. (2000) *Neural Correlates of Consciousness: Empirical and Conceptual Questions* Metzinger, T. (Ed.), What is a Neural Correlate of Consciousness? (MIT Press).

Churchland, P. M. (1988). Matter and consciousness : a contemporary introduction to the philosophy of mind Rev. ed. (MIT Press Cambridge, Mass.).

Crick, F. (1994). The Astonishing hypothesis : the scientific search for the soul. (Simon & Schuster London).

Crutchfield, J. P. (1994). The calculi of emergence: computation, dynamics and induction. *Physica D,* 75, pp. 11-54.

Dawkins, R. (2006). *The Blind Watchmaker.* (Penguin London).

Denton, D. (2005). The Primordial Emotions: The dawning of consciousness. (Oxford University Press).

Edelman, G. M. and Tononi, G. (2000). *A universe of consciousness : how matter becomes imagination* 1st ed. (Basic Books New York, NY).

Einstein, A. (1950). *Out of my later years.* (Philosophical Library New York).

Eisenberg, R. S. (1999). From structure to function in open ionic channels. *Journal of Membrane Biology,* 171, pp. 1-24.

Feibleman, J. K. (1954). Theory of Integrative Levels. *The British Journal for the Philosophy of Science,* 5, pp. 59-66.

Feigl, H. (1958) *Concepts, Theories, and the Mind-Body Problem* Feigl, H.and M. Scriven, *et al.* (Eds.), The "Mental" and the "Physical". (University of Minnesota Press).

Frohlich, F. and McCormick, D. A. (2010). Endogenous Electric Fields May Guide Neocortical Network Activity. *NEURON, 67*, pp. 129-143.

Gamez, D. (2008). Progress in machine consciousness. *Consciousness and Cognition, 17*, pp. 887-910.

Gelertner, D. (2007). Artificial Intelligence is Lost in the Woods. In *Technology Review*: MIT.

Gould, S. J. and Eldredge, N. (1972). Punctuated equilibria: an alternative to phyletic gradualism. *Models in paleobiology*, pp. 82-115.

Haikonen, P. O. (2012). *Consciousness and Robot Sentience*. (World Scientific).

Hales, C. G. (2009a). Dual Aspect Science. *Journal of Consciousness Studies, 16*, pp. 30-73.

Hales, C. G. (2009b). An empirical framework for objective testing for P-consciousness in an artificial agent. *The Open Artificial Intelligence Journal, 3*, pp. 1-15.

Harnad, S. (1991). Other bodies, other minds: A machine incarnation of an old philosophical problem. *Minds and Machines, 1*, pp. 43-54.

Hess, D. J. (1997). *Science studies : an advanced introduction*. (New York University Press New York).

Hofstadter, D. R. (1980). *Gödel, Escher, Bach : an eternal golden braid*. (Penguin Harmondsworth).

Holmes, N. (2003). Artificial intelligence: Arrogance or ignorance? *Computer, 36*, pp. 120-+.

Hopgood, A. A. (2003). Artificial intelligence: Hype or reality? *Computer, 36*, pp. 24-+.

Horgan, J. (1992). The New Challenges. *Scientific American, 267*, pp. 16-&.

Hume, D. and Steinberg, E. (1993). An enquiry concerning human understanding ; [with] A letter from a gentleman to his friend in Edinburgh ; [and] An abstract of a Treatise of human nature 2nd ed. (Hackett Pub. Co. Indianapolis).

Izhikevich, E. M. (2007). Dynamical systems in neuroscience : the geometry of excitability and bursting. (MIT Press Cambridge, Mass. ; London).

James, W. (1890). *The principles of psychology*. (Macmillan. London).

Kandel, E. R. and Schwartz, J. H. (2000). *Principles of neural science* 4th , International ed. (McGraw-Hill New York).

Kirk, R. (2012). Zombies. In Zalta, E. N. (Ed.), *The Stanford Encyclopedia of Philosophy* (Summer 2012 ed.).

Kitcher, P. (1993). The Advancement of Science: Science without legend, objectivity without illusions. (Oxford University Press New York).

Koch, C. and Tononi, G. (2011). A Test for Consciousness. *Scientific American, 304*, pp. 44-47.

Koestler, A. (1967). *The ghost in the machine*. (Hutchinson London).

Koestler, A. (1978). *Janus : a summing up* 1st American ed. ed. (Random House New York :).

Kuhn, T. S. and Hacking, I. (2012). *The structure of scientific revolutions* 4th ed. (University of Chicago Press Chicago).

Leake, D. (2006). Fifty years of artificial intelligence research. *Ai Magazine,* 27, pp. 3-3.

Lehar, S. (2003). Gestalt isomorphism and the primacy of subjective conscious experience: A Gestalt Bubble model. *Behavioral and Brain Sciences,* 26, pp. 375-443.

Lewes, G. H. (1879). *Problems of life and mind.* (Trubner London).

Lewis, C. I. (1929). Mind and the world-order; outline of a theory of knowledge. (Dover Publications New York).

Macdonald, C. (1989). *Mind-body identity theories.* (Routledge London ; New York).

Mach, E. (1897). *Contributions to the analysis of the sensations.* (Open Court Pub. Co. La Salle, Ill.).

Magnani, L. (2009). Abductive cognition: The epistemological and eco-cognitive dimensions of hypothetical reasoning Vol. 3. (Springer).

McCrone, J. J. (2004). How do you persist when your molecules don't? In Baars, B. and A. Revonsuo, *et al.* (Eds.), *Science & Consciousness Review* (SCR 2004, June, No. 1 ed.).

Mead, C. (1989). *Analog VLSI and neural systems.* (Addison-Wesley Reading, Mass).

Mead, C. (1990). Neuromorphic electronic systems. *Proceedings of the Ieee,* 78, pp. 1629-1636.

Mill, J. S. (1930). A system of logic, ratiocinative and inductive. Being a connected view of the principles of evidence and the methods of scientific investigation. (Longmans Green London).

Miller, D. (2005). *Out of Error.* (Ashgate).

Mitchell, M. and Crutchfield, J. P., *et al.* (1994). Evolving cellular-automata to perform computations - mechanisms and impediments. *Physica D,* 75, pp. 361-391.

Mithen, S. (2002) *The cognitive basis of science* Carruthers, P.and S. P. Stich, *et al.* (Eds.), Human evolution and the cognitive basis of science. (Cambridge University Press, New York), pp. 23-40.

Moor, J. (2006). The Dartmouth College Artificial Intelligence Conference: The next fifty years. *Ai Magazine,* 27, pp. 87-91.

Moyer, M. (2011). Watson Looks for Work. *Scientific American Magazine,* 304, pp. 19-19.

Mullins, J. (2005). Whatever happened to machines that think? *New Scientist,* 186, pp. 32-37.

Munakata, T. (1996). Thoughts on Deep Blue vs. Kasparov. *Communications of the ACM,* 39, pp. 91-92.

Nagel, E. (1961) The structure of science : problems in the logic of scientific explanation Patterns of Scientific Explanation. (Routledge, London).

Nagel, E. and Newman, J. R., *et al.* (2002). *Gödel's proof* Rev. ed. (New York University Press New York).

Nagel, T. (1974). What is it like to be a bat? *The Philosophical Review*, pp. 435-450.

Ne'eman, Y. and Eizenberg, E. (1995). Membranes and other Extendons ("p-BRANES"). Classical and Quantum Mechanics of Extended Geometrical Objects Vol. 39. (World Scientific).

Polanyi, M. (1967). *The tacit dimension.* (Routledge & K. Paul London,).

Popper, K. R. (1999). *All life is problem solving.* (Routledge London ; New York).

Prigogine, I. and Stengers, I. (1985). *Order out of chaos : man's new dialogue with nature* Flamingo ed. (Fontana Paperbacks London).

Prigogine, I. and Stengers, I. (1997). *The end of certainty : time, chaos, and the new laws of nature* 1st Free Press ed. (Free Press New York).

Rasmussen, S. and Baas, N. A., *et al.* (2008) *Emergence : contemporary readings in philosophy and science* Bedau, M.and P. Humphreys (Eds.), Ansatz for Dynamical Hierarchies. (MIT Press, Cambridge, Mass.), pp. 269-286.

Reddy, R. (1996). The challenge of artificial intelligence. *Computer,* 29, pp. 86-&.

Reimann, Michael W. and Anastassiou, Costas A., *et al.* (2013). A Biophysically Detailed Model of Neocortical Local Field Potentials Predicts the Critical Role of Active Membrane Currents. *NEURON,* 79, pp. 375-390.

Robbins, P. and Aydede, M. (2009). The Cambridge Handbook of Situated Cognition. In (pp. xi, 520 p.). Cambridge; New York: Cambridge University Press.

Rogoff, B. and Chavajay, P., *et al.* (1993). Questioning assumptions about culture and individuals. *Behavioral and Brain Sciences,* 16, pp. 533-533.

Rohen, J. W. and Yokochi, C., *et al.* (1988). *Color atlas of anatomy : a photographic study of the human body* 2nd ed. (Igaku-Shoin New York ; Tokyo).

Rovelli, C. (2006). Graviton propagator from background-independent quantum gravity. *Physical Review Letters,* 97, pp. -.

Sen, A. (1998). An introduction to non-peturbative string theory. In (pp. 130).

Shapiro, L. A. (2013). Dynamics and Cognition. *Minds and Machines,* 23, pp. 353-375.

Shi, Z. Z. and Zheng, N. N. (2006). Progress and challenge of artificial intelligence. *Journal of Computer Science and Technology,* 21, pp. 810-822.

Smart, J. J. C. (2004). The Identity Theory of Mind. In Zalta, E. N. (Ed.), *The Stanford Encyclopedia of Philosophy* (Vol. Fall 2004 Edition).

Swarup, A. (2006). Sights set on quantum froth. *New Scientist,* 189, pp. 18-18.

Tomasello, M. and Kruger, A. C., *et al.* (1993). Cultural learning. *Behavioral and Brain Sciences,* 16, pp. 495-511.

Tye, M. (2009). Qualia. In Zalta, E. N. (Ed.), *The Stanford Encyclopedia of Philosophy* (Summer 2009 ed.).

Varela, F. J. and Thompson, E., *et al.* (1991). *The embodied mind : cognitive science and human experience.* (MIT Press Cambridge, Mass.).

Varsavsky, A. (2009). Epileptic Seizures and the EEG: Measurement, Models, Detection and Prediction. University of Melbourne, Melbourne, Australia.

Velmans, M. and Schneider, S. (2007). The Blackwell companion to consciousness. In. Malden, MA ; Oxford: Blackwell Publishing.

Wallace, B. A. (2000). The taboo of subjectivity : toward a new science of consciousness. (Oxford University Press New York).

Wang, C. L. and Ahmed, P. K. (2003). Organisational learning: a critical review. *Learning Organization, The,* 10, pp. 8-17.

Weinert, F. (2013). Lines of Descent: Kuhn and Beyond. *Foundations of Science*, pp. 1-22.

Wertheim, M. (2011). Physics on the fringe : smoke rings, circlons, and alternative theories of everything. (Walker New York).

Wolfram, S. (2002). *A new kind of science.* (Wolfram Media Champaign, IL).

Zeman, A. (2001). Consciousness. *Brain,* 124, pp. 1263-1289.

Index

Printed in the United States
By Bookmasters